Barry J. Nalebuff graduated from Oxford University and is now the Milton Steinbach Professor at Yale School of Management. He is co-author of *Thinking Strategically* (with Avinash Dixit). He has consulted to American Express, Bell Atlantic, Citibank, Corning, General Re, IC-O, Merck, McKinsey, Procter & Gamble, and RTZ, and is a principal of the Law and Economics Consulting Group.

Adam M. Brandenburger graduated from Cambridge University and is now a professor at the Harvard Business School. He is the author of many of Harvard's bestselling strategy cases. He has worked with Ciba-Geigy, Fidelity Investments, Honeywell, KPMG Peat Marwick, Merck, and Northwestern Life Insurance.

They can be reached at barry.nalebuff@yale.edu and abrandenburger@hbs.edu and at the home page, http://mayet.som.yale.edu/coopetition

Co-opetition

- 1. *A revolutionary mindset that combines competition and cooperation*
- 2. The Game Theory strategy that's changing the game of business

PROFILE BOOKS

Published by
PROFILE BOOKS LTD
58A Hatton Garden
London EC1N 8LX
www.profilebooks.co.uk

First published in Great Britain by HarperCollinsBusiness 1996
First published in the USA by Currency, Doubleday 1996

3 5 7 9 10 8 6 4 2

Printed and bound in Great Britain by
Bookmarque Ltd, Croydon, Surrey

The moral right of the authors has been asserted.

A CIP catalogue record for this book is available from the
British Library.

ISBN 1 86197 507 4

To our parents:
Marcia and Edward, Ennis and Frank

Contents

Production Notes

The combination of theory and practice you're about to read has taken many years to develop. Throughout these years we have received enormous help from people in the academic and business worlds and from friends and family. Our personal and intellectual debts are numerous and large.

We were each lucky enough to have unusually brilliant and inspirational teachers to initiate us into the field of game theory. Louis Makowski showed Adam the value of looking at everything from unusual angles. Bob Aumann taught Adam to think hard in order to make things simple. Bob Solow, a disarmingly modest Nobel laureate, taught Barry the power of asking the right questions. The exuberant intellectual curiosity of Joe Stiglitz and Richard Zeckhauser inspired Barry to explore the wider applications of game theory.

Over the years, our research has been supported by the Harkness Foundation, the Harvard Business School Division of Research, the Harvard Society of Fellows, the National Science Foundation, the Pew Charitable Trust, the Rhodes Trust, the Alfred P. Sloan Foundation, and the Yale School of Management. We are extremely grateful for the generosity of all these institutions. Their funding allowed us to do the basic research that led to this book.

At Harvard Business School, former dean John McArthur and Mike Porter have been a constant support to Adam in his work. Anita McGahan, Dick Rosenbloom, Gus Stuart, and David Yoffie are among Adam's colleagues who have been enthusiastic supporters and keen critics. In fact, this book wouldn't have happened without Gus. He is a co-inventor of some of the key concepts that structure our approach.

At Princeton, Avinash Dixit got Barry started on writing books, coauthoring *Thinking Strategically: The Competitive Edge in Business, Politics, and Everyday Life*. Former dean Mike Levine brought Barry to the Yale School of Management and encouraged him to create a course in game theory. Sharon Oster guided Barry in the transition to business strategy.

We have been fortunate to regularly teach wonderful students at Harvard and Yale – and we learned as we taught. Our early courses on game theory and business were what you might call successful failures. We didn't yet have the synthesis of theory and practice. The weaknesses of these early courses taught us a great deal about what was missing from our understanding. This book is the direct result of those early efforts. We thank all of our early students for bearing with us during this period of experimentation and learning.

As we developed and extended our new synthesis, we drew heavily on research provided by our students and assistants over the previous years. The students who worked on cases that appear in this book include: Greg Camp, Greg Chin, David Cowan, Michael Maples, Anna Minto, Richard Malloy, David Myers, Paul Sullivan, Bartley Troyer, Michael Tuchen, and Peter Wetenhall. The research assistants who supplied essential material include: Christine Del Ballo, Paul Barese, Monique Burnett, Maryellen Costello, Brad Ipsan, Julia Kou, Fiona Murray, Troy Paredes, Adam Raviv, Deepak Sinha, and Geoff Verter. We'll miss Troy's 3:00 A.M. voice mails – and his surprise when we answered the phone.

We appreciate the opportunities the following people have given us to try out these ideas in the field: Ken Chenault and Andy Wing at American Express, Charles Freeman at Chemical Bank, Robert Clement and Lynn Stair at Citibank, Jason Walsh and Jim Cooke at Corning, Ron Ferguson and T. Hoffman at General Re, Andy Shearer at KPMG Peat Marwick, Geoff Porges at Merck, Mike Keller (formerly) at Northwestern Life Insurance, Lydia Marshall at Sallie Mae, Mark Myers at Xerox. Bill Roughton at Bell Atlantic provided a unique opportunity to work on the Federal Communications Commission auction of personal communication service spectrum. Bill Barnett at McKinsey and Co. challenged us to make game theory relevant, gave us the chance to work with his clients, and was invaluable in helping us bridge the gap between theory and practice. It would be hard to give enough thanks to our corporate clients for the enormous amount they have taught us. In addition, the constant feedback from executive education programs and seminars helped us shape this book.

We are indebted to the *Harvard Business Review* for promoting our work and improving it along the way. The process of preparing an article proved extremely valuable. This was largely because of the encouragement and critical feedback we received from Joan Magretta, Nancy Nichols, Sharon Slodki, and Nan Stone.

When we turned to write this book, Loretta Barrett helped get us started. Helen Rees, our North American agent, and Linda Michaels, our foreign rights agent, continue to amaze us with their insights and skill. The enormous enthusiasm, confidence, and – yes – patience of Bill Thomas, our editor at Doubleday, gave the project an enormous boost throughout. Harriet Rubin, of Currency/Doubleday, provided wonderful criticism, always making sure that we had enough 'trope.'

Scott Borg, novelist and cultural historian, was a brilliant help in making portions of the text clearer and more readable. He pushed us where we needed to be pushed, and pulled us forward with his insights and logic.

Early on, we discovered the skills of Rena Henderson, who does brilliant, high-speed manuscript editing from her Monterey, California, company, As the Word Turns. Never have we felt that someone knows us so well who's never met us.

At every stage in the writing of this book, we benefited enormously from the many people who read and criticized our various drafts. Academic colleagues who provided informed critiques included: Bharat Anand, Sushil Bikhchandani, Joe Bower, Jeremy Bulow, David Collis, Ken Corts, John Geanakoplos, Oscar Hauptman, Bob Kennedy, Tarun Khanna, Elon Kohlberg, Ben Polak, Julio Rotemberg, Roni Shachar, Carl Shapiro, Debra Spar, and Elizabeth Teisberg.

Current and former students who gave us valuable feedback on drafts of the book included: Terry Burnham, Putnam Coes, Amy Guggenheim, Roger Hallowell, Walter Kümmerle, Jon Levin, Matt Littlejohn, Amir Makov, Andrew McAfee, Robin Mendelson, Roslyn Romberg, Ed Simnett, Hunt Stookey, Don Sull, and Mike Troiano. Eric Muller's comments were so complete, they could almost have been published as a separate work. Old friends and new friends who offered insights from their businesses include: Christine Bucklin, Jim Cooke, Bob Cozzi, Bob Davoli, T. Hoffman, Mark Kaminsky, Jeff Keisler, John MacBain, Frank Murphy, Elizabeth Shackleford, Lenny Stern, Patrick Viguerie, Jason Walsh, Mary Westheimer, and Evan Wittenberg.

Close friends and family members were especially patient and helpful in this project. Diane Rubin, Jim Cook, Lionel Fray, Larry Hilibrand, Warren Spector, and Bob Taylor reviewed early drafts. Above and beyond any call of duty or friendship, Ken French, Stephen Scher, and John Lapides went over every page with us. Their unstinting efforts are deeply, deeply appreciated. Ennis Brandenburger went over all the material with a meticulous mother's eye. Back at home, our greatest debt is to Barbara Rifkind and Helen Kauder, who, along with their critical reading of this book, are great supporters of everything we do.

We have been fortunate to be helped in this project by so many people in all different walks of life — CEOs, human resource managers, marketers, small-business owners, lawyers, entrepreneurs, nonprofit managers, academics, business school students, undergraduate students, artists, and mothers. We hope that we have succeeded in writing a book that will be useful to all those who have so generously helped us (and to many other people, too).

ADAM BRANDENBURGER
BARRY NALEBUFF
January 1996

Foreword to the Paperback Edition

There's nothing so practical as a good theory. A good theory confirms the conventional wisdom that 'less is more.' A good theory does less because it doesn't give answers. At the same time, it does a lot more because it helps people organize what they know and uncover what they don't know. A good theory gives people the tools to discover what is best for them. That was our goal in writing Co-opetition.

Co-opetition offers a theory of value. It's a book about creating value and capturing value. There's a fundamental duality here: whereas creating value is an inherently cooperative process, capturing value is inherently competitive. To create value, people can't act in isolation. They have to recognize their interdependence. To create value, a business must align itself with customers, suppliers, employees, and many others. That's the way to develop new markets and expand existing ones.

But along with creating a pie, there's the issue of dividing it up. This is competition. Just as businesses compete with one another for market share, customers and suppliers also are looking out for their slice of the pie.

Creating value that you can capture is the central theme in Co-opetition.

Obviously the best way to do this will be different for different businesses. But one strategy that Co-opetition emphasizes is working with what we term 'complementors.' A complementor is the opposite of a competitor. It's someone who makes your products and services more, rather than less, valuable. Not surprisingly, the complementor concept is especially relevant to the builders of the Information Economy. Hardware needs software, and the Internet needs high-speed phone lines. No one, alone, can build the infrastructure for the new economy. It's a whole new system made up of many complementary parts.

Thinking about the new economy, we've realized that there's a deeper connection here. The connection is through one of the great intellectual figures of this century, John von Neumann.

John von Neumann – mathematician, genius, polymath – died in 1957, well before he could see the emergence of the Information Age he helped to create. He was a co-inventor of the modern computer architecture – today's programmable computer. He was also responsible for pioneering work on self-reproducing systems, presaging the discovery of DNA. And, together with economist Oskar Morgenstern, he was the inventor of game theory. His theory of games provides a model of the pie, and how it gets divided up. We rely on these insights throughout Co-opetition.

Game theory is a different way of looking at the world. Conventional economics takes the structure of markets as fixed. People are thought of as simple stimulus-response machines. Sellers and buyers assume that products and prices are fixed, and they optimize production and consumption accordingly. Conventional economics has its place in describing the operation of established, mature markets, but it doesn't capture people's creativity in finding new ways to interact with one another.

In game theory, nothing is fixed. The economy is dynamic and evolving. The players create new markets and take on multiple roles. They innovate. No one takes products or prices as given. If this sounds like the free-form and rapidly transforming marketplace, that's precisely why game theory may be the kernel of a new economics for the new economy. And that is why we see Co-opetition as a book for the Information Age.

Traditionally, a book has been a static, one-way medium. Fortunately, that's changing. Like many authors, we've been able to use the Internet to make our interaction with readers more dynamic and interactive. On the Co-opetition home page, you'll find updates, articles, some interactive exercises, overheads, audio, and a convenient way to email us. Since the book first came out, we've learned a great deal from a great many people about how and where they've been putting Co-opetition to work. Our heartfelt thanks to all of them. We hope you'll share your reactions with us, too.

ADAM BRANDENBURGER (abrandenburger@hbs.edu)
BARRY NALEBUFF (barry.nalebuff@yale.edu)
April 1997
Home page: http://mayet.som.yale.edu/coopetition

Part 1

The Game of Business

1 War and Peace

'**B**usiness is War.' The traditional language of business certainly makes it sound that way: outsmarting the competition, capturing market share, making a killing, fighting brands, beating up suppliers, locking up customers.[1] Under business-as-war, there are the victors and the vanquished. The ultimate win-lose view of the world comes from author Gore Vidal:

> ### It is not enough to succeed. Others must fail.

But the way people talk about business today, you wouldn't think so. You have to listen to customers, work with suppliers, create teams, establish strategic partnerships – even with competitors. That doesn't sound like war. Besides, there are few victors when business is conducted as war. The typical result of a price war is surrendered profits all around. Just look at the US airline industry: it lost more money in the price wars of 1990–93 than it had previously made in all the time since Orville and Wilbur Wright.[2]

The antithesis to Gore Vidal's worldview comes from Bernard Baruch, a leading banker and financier for much of this century:

> ### You don't have to blow out the other fellow's light to let your own shine.

Though less famous today than Gore Vidal, Baruch made a whole lot more money. More often than not, we'll follow Baruch's advice in this book.

In fact, most businesses succeed only if others also succeed. The demand for Intel chips increases when Microsoft creates more powerful software. Microsoft software becomes more valuable when Intel produces faster chips. It's mutual success rather than mutual destruction. It's win-win. The cold war is over and along with it the old assumptions about competition.

So,

'Business is Peace'?

That doesn't sound quite right, either. We still see battles with competitors over market share, fights with suppliers over cost, and conflicts with customers over price. And the success of Intel and Microsoft hasn't exactly helped Apple Computer. So if business isn't war and it isn't peace, what is it?

A new mindset

Business is cooperation when it comes to creating a pie and competition when it comes to dividing it up. In other words, business is War *and* Peace. But it's not Tolstoy – endless cycles of war followed by peace followed by war. It's simultaneously war and peace. As Ray Noorda, founder of the networking software company Novell, explains: 'You have to compete and cooperate at the same time.'[3] The combination makes for a more dynamic relationship than the words 'competition' and 'cooperation' suggest individually. This is why we've adopted Noorda's word *co-opetition*, and made it the title of our book.

What's the manual for co-opetition? It's not Leadership Secrets of Attila the Hun.[4] Nor is it Leadership Secrets of St Francis of Assisi. You can compete without having to kill the opposition. If fighting to the death destroys the pie, there'll be nothing left to capture – that's lose-lose. By the same token, you can cooperate without having to ignore your self-interest. After all, it isn't smart to create a pie you can't capture – that's lose-win.

The goal is to do well for yourself. Sometimes that comes at the expense of others, sometimes not. In this book, we'll discuss business as a game, but not a game like sports, poker, or chess, which must be win-lose. In business, your success doesn't require others to fail – there can be multiple winners. Throughout the book, you'll see many examples of this. In the spirit of co-opetition, we'll present some cases where win-lose is the most effective approach and others where win-win is most effective. We'll discuss situations where defeating your competitors is the best course, and present other situations where the best plan benefits several players, including competitors.

Putting co-opetition into practice requires hardheaded thinking. It's not enough to be sensitized to the possibilities of cooperation and win-win strategies. You need a framework to think through the dollars-and-cents consequences of cooperation *and* of competition.

Game theory

To find a way of bringing together competition and cooperation, we turn to game theory. Game theory has the potential to revolutionize the way people think about business. This is because the fundamental ideas of game theory are so powerful, and because business offers so many opportunities for applying them.

There has been a growing recognition that game theory is a crucial tool for understanding the modern business world. In 1994 three pioneers in game theory – John Nash, John Harsanyi, and Reinhart Selten – were awarded a Nobel Prize. At the same time, the Federal Communications Commission was using game theory to help it design a $7-billion auction of radio spectrum for personal communication services. (Naturally, the bidders used game theory too.) Even as we write, the leading management consulting firms are introducing game theory into their strategy practices.

The field of game theory dates back to the early days of World War II, when British naval forces playing cat and mouse with German submarines needed to understand the game better so that they could win it more often.[5] They discovered that the right moves weren't the ones pilots and sea captains were making intuitively. By applying concepts later known as game theory, the British improved their hit rate enormously. Their success against submarines led them to apply game theory to many other war activities. Thus, game theory was proven in practical life-and-death situations before it was actually laid out on paper as a systematic theory.

The classic theoretical formulation came soon after, in 1944, when mathematical genius John von Neumann and economist Oskar Morgenstern published their book *Theory of Games and Economic Behavior*. This brilliant, but highly abstract, work was immediately heralded as one of the greatest scientific achievements of the century. It led to large numbers of technical papers in the fields of economics, politics, military strategy, law, computer science, and even evolutionary biology. In each of these fields, game theory has resulted in major discoveries. Now game theory is transforming the field of business strategy.

Game theory makes it possible to move beyond overly simple ideas of competition and cooperation to reach a vision of co-opetition more suited to the opportunities of our time. To many, this will come as a surprise. The image game theory often conjures up is business-as-war. That's to be expected, since the field was born during World War II and grew up during the Cold War. The mentality was one of winners and losers – the zero-sum game, even the zero-sum society.[6] But that's only half the subject. Contemporary game theory applies just as well to positive-sum – or win-win – games. The real value of game theory for business comes when the full theory is put into practice: when game theory is applied to the interplay between competition *and* cooperation.

What are the essential characteristics of game theory as applied to business? What are its special virtues? How is it different from a host of other management tools?

What game theory has to offer

Game theory focuses directly on the most pressing issue of all: finding the right strategies and making the right decisions. There are many valuable books on how to create a management environment conducive to making the right decisions. There are also valuable books on how to build organizations effective at carrying out decisions once they're made. But there's still a great need for guidance in identifying the right strategy to begin with. This is what game theory provides. It goes right for the crux of things, showing you in strategic terms what is the best thing to do.

Game theory is particularly effective when there are many interdependent factors and no decision can be made in isolation from a host of other decisions. Business today is conducted in a world of bewildering complexity. Factors you might not even think to ask about can determine your success or failure. Even if you identify all the relevant factors, anything that changes one is likely to affect many others. Amid all this complexity, game theory breaks down the game into its key components. It helps you see what's going on and what to do about it.

Game theory is an especially valuable tool to share with others in your organization. The clear and explicit principles of game theory make it easier to explain the reasoning behind a proposed strategy. It gives you and your colleagues a common language for discussing alternatives. By letting others in on the process you've used to reach a strategic decision, game theory helps you build a consensus.

Such techniques for sharing strategic thinking are increasingly needed at all levels of business. Decision making is becoming more complex and more decentralized. Rapid changes in markets and technology require rapid, strategically informed responses. Hence, the number of people in a company who will benefit from applying game theory is growing greater all the time.

Game theory is an approach you can expand and build on. It's not a particular prescription suited to a particular moment in business history. It's not a rule of thumb that stops working when conditions change. It's a way of thinking that survives changing business environments.

In many cases, game theory can suggest options that otherwise might never have been considered. This is a consequence of game theory's systematic approach. By presenting a more complete picture of each business situation, game theory makes it possible to see aspects of the situation that would otherwise have been ignored. In these neglected aspects, some of the greatest opportunities for business strategy are to be found.

What you'll find in this book

We approach game theory mainly through real-life stories, involving characters and companies you'll recognize. These stories tell of businesses competing and cooperating, succeeding and failing, sometimes with surprising outcomes. Some are war stories, others are peace stories. In both cases, they are more than anecdotes. We use game theory to explain the successes and failures. Each story is a case study accompanied by a full analysis of the principles involved. We interweave the stories with theory, and summarize the lessons in the form of checklists. This way, our analysis becomes more than descriptive. It becomes prescriptive, too. When you understand why a strategy worked – or didn't – you can apply the lesson to other situations.

The numerous case studies have other functions as well. They're not just a device for making the subject more entertaining or for showing how our concepts work in practice. They serve as an ongoing test of our theories. We're skeptics, and we want you to be skeptical, too. We don't want you to take what we say on trust. Our goal is to give you enough evidence through case studies to accept or challenge our conclusions. After you've seen game theory applied to large numbers of cases, you'll discover its power, get a feel for how it works, and learn to apply it yourself.

Despite the current surge of interest in applying game theory to business, this is still a very new approach. Much of the terminology is new. In fact, some of the key terms were actually coined during the writing of this book. Even terms that seem familiar take on a new meaning in the context of game theory. Like any theory offering a new perspective, it requires some patience in the beginning. But if our explanations are successful, the new concepts will soon become so much a part of your thinking that you'll wonder how you ever managed without them.

How this book is organized

Part 1, consisting of three chapters, outlines the game of business. It introduces all the basic concepts and shows how they fit together. The present chapter is intended to serve as an orientation session, a kind of advance briefing on where this book will take you.

Chapter 2 describes all the players and analyzes the elements of competition and cooperation among them. To make this clear, we construct a map for the game of business. We call this the Value Net. It's a diagram that serves as a visual representation of the game of business. The Value Net locates all the various players relative to one another, and identifies the interdependencies among them. It's particularly useful for pointing out the ways a relationship between players can combine competition and cooperation.

Chapter 3 introduces game theory. We explain how this academic discipline applies to the real world of business. Using detailed examples, we discuss what happens when games are played out. In the process, we make game theory accessible by taking the essential principles and stating them in a simple and clear fashion that requires no mathematics or abstract theory.

Our account of game theory identifies five basic elements of any game: *Players*, *Added values*, *Rules*, *Tactics*, and *Scope* – PARTS, for short. These become our touchstones for the rest of the book. Along with the Value Net, they provide the central conceptual scheme for applying game theory to business.

Part 2, the remainder of the book, consists of separate chapters on each of the five elements of a game. We describe each element in detail and what significance each has for your business. Archimedes said that given a proper lever, he could move the world. These are the five levers for moving the world of business.

Changing the game

This is where the biggest payoff comes. We said business is different from other games because it allows more than one winner. But business is also different in another fundamental way: the game doesn't stand still. All the elements in the game of business are constantly changing; nothing is fixed. This is not just by chance. While football, poker, and chess have ultimate ruling bodies – the NFL or FIFA, Hoyle, and the Chess Federation – business doesn't.[7] People are free to change the game of business to their benefit. And they do.

Why change the game? An old Chinese proverb explains: if you continue on the course you're headed, that's where you'll end up. Sometimes that's good, sometimes not. You can play the game extremely well, and still fare terribly. That's because you're playing the wrong game: you need to change it. Even a good game can be made into a better one. Real success comes from actively shaping the game you play – from making the game you want, not taking the game you find.

How do you change the game? You may well have been doing this instinctively. But game theory provides a systematic method. To change the game, you have to change one or more of the five elements: you change the PARTS. Each component we discuss is a powerful tool for transforming the game into a different one. This is where game theory finds its greatest opportunities: in *changing* the game. Changing not just *the way* you play, but *the game* you play.

2 Co-opetition

If business is a game, who are the players and what are their roles? There are customers and suppliers, of course; you wouldn't be in business without them. And, naturally, there are competitors. Is that it? No, not quite. There's one more, often overlooked but equally important group of players – those who provide complementary rather than competing products and services. That's where we'll begin this chapter. We'll see how complements can make the difference between business success and failure.

1 Thinking Complements

The classic example of complements is computer hardware and software. Faster hardware prompts people to upgrade to more powerful software, and more powerful software motivates people to buy faster hardware. For example, Windows 95 is far more valuable on a Pentium-powered machine than on a 486 machine. Likewise, a Pentium chip is far more valuable to someone who has Windows 95 than to someone who doesn't.

Though the idea of complements may be most apparent in the context of hardware and software, the principle is universal. A complement to one product or service is any other product or service that makes the first one more attractive. Hot dogs and mustard, cars and auto loans, televisions and videocassette recorders, television shows and *TV Guide*, fax machines and phone lines, digital cameras and colour printers, catalogs and overnight delivery services, red wine and dry cleaners, Siskel and Ebert. These are just some of the many, many examples of complementary products and services.

Let's take a closer look at the complements to cars. An obvious one is paved roads. Having built a better mousetrap, the fledgling auto industry didn't leave it to others to make a beaten path to its door. While it couldn't pave all the roads itself, it got many started. In 1913 General Motors, Hudson, Packard, and Willys-Overland, together with Goodyear tires and Prest-O-Lite headlights, set up the Lincoln Highway Association to catalyze development of America's first coast-to-coast highway.[1] The association built 'seedling miles' along the proposed transcontinental route. People saw the feasibility and value of paved roads and lobbied the government to fill in the gaps. In 1916 the federal government committed its first dollars to building roads; by 1922 the first five transcontinental highways, including the Lincoln, had been completed.

Today there are plenty of roads, but money can still be scarce. Cars, especially new ones, are expensive, so if customers find it hard to borrow, they may find it hard to buy a new car. Thus, banks and credit unions complement Ford and General Motors. But auto financing has not *always* been accessible. That's why General Motors created General Motors Acceptance Corporation back in 1919 and Ford Motors formed Ford Motor Credit in 1959. It doesn't really matter who provides the financing – banks, credit unions, or the automobile credit companies themselves. More money in this market leads to lower interest rates. Better and cheaper access to credit enables people to buy more cars – and that helps Ford and GM. The flip side is also true: selling cars helps Ford and GM sell loans. Over the last decade, Ford has actually earned more money making loans than making cars.

Auto insurance is a complement to cars because, without insurance, people might not be willing to risk investing $20,000 or more in a new car. Just as carmakers have made auto loans more affordable, perhaps they could help make auto insurance more affordable. This would be particularly valuable to first-time buyers, who often face prohibitively high insurance rates.

Complements are always reciprocal. Just as auto insurance complements new cars, new cars complement auto insurance. The more new cars people buy, the more insurance they buy, especially collision and theft insurance. Thus, auto insurance companies might want to use their expertise and clout to help their customers get a better price on new cars. We'll come back to the subject of cars and auto insurance later in the book.

Suppliers to the car industry haven't forgotten complements, either. Until tire manufacturers figure out a way to add a fifth wheel to a car, there's really only one way for them to boost sales, and that's to whet people's appetite to drive. That's why the French tire maker Michelin sells the Michelin guidebooks. These guidebooks don't give just the shortest route, they make sure to point out the longer scenic routes as well. The Michelin guide makes getting there at least half the fun. It encourages travelers to keep moving, to keep wearing down those tires. There's always another town to see, another interesting detour to make. The Michelin guide not only helps sell more tires, it's also a profitable business in its own right. It dominates the guidebook market in France and is making inroads in the rest of Europe.

The used-car market also benefits when people pay attention to complements. For proof, look to John and Louise MacBain, publishers of *La Centrale des Particuliers*, a Paris weekly specializing in advertisements for used cars. They have found people who will provide the complementary services their readers want – auto insurance, financing, and mechanical warranties. And the MacBains have prenegotiated very favorable rates for their readers by giving the providers a prominent listing in the magazine and use of the *La Centrale* brand name. The MacBains go even further, selling some complements themselves. Both readers and advertisers want to know average transaction prices and the average time on the market for each make, model, and year. Through France's Minitel online service, the MacBains make this information available for a fee.[2] By paying attention to complements, the MacBains have ensured that there is no competition for *La Centrale*. They've taken their idea on the road, changing the way used cars are sold in Canada, Hungary, Poland, Sweden, Thailand, the United States, and other countries.

The complements mindset also helps explain why some businesses fail. Alfa Romeo and Fiat had trouble selling their cars in the United States, because people knew they'd have trouble finding spare parts and qualified mechanics. Both have exited the US market. The Sony Betamax videocassette recorder, though technically superior to the VHS in some respects, was ultimately undone by the lack of rental movies in the Betamax format. In many cities, downtown shopping has lost out to suburban malls because of a lack of convenient parking. If these enterprises had provided the necessary complements, they might have fared much better.

The problem of missing complements is multiplied a thousand

times over in the case of a new economy. This is the situation in much of the third world and in many of the former communist countries. There the fate of everything – not just the company or industry but often the whole country – depends on complements. One industry will need complementary industries so it can get going, but those complementary industries will need the first industry so they can get going. It's a chicken-and-egg situation everywhere you look. Everything has to happen all together, or nothing might happen at all. That's why some developing economies take off while others stall.

Thinking complements is a different way of thinking about business. It's about finding ways to make the pie bigger rather than fighting with competitors over a fixed pie. To benefit from this insight, think about how to expand the pie by developing new complements or making existing complements more affordable.

Intel is the ultimate competitor. Andy Grove, the CEO, is known for saying: 'Only the paranoid survive.'[3] But competitors aren't the only thing on Grove's mind; Intel is also on the lookout for complements.

Inside Intel We started this chapter by explaining how Microsoft benefits when Intel develops a faster chip and how Intel benefits when Microsoft pushes forward in software development. But from Intel's perspective, Microsoft doesn't push hard enough. According to Andy Grove: 'Microsoft doesn't share the same sense of urgency [to come up with an improved PC]. The typical PC doesn't push the limits of our microprocessors . . . It's simply not as good as it should be, and that's not good for our customers.'[4]

If software applications don't push the limits of existing microprocessor chips, then Grove has to find something else that will. Otherwise, his customers won't feel the continued need to upgrade. If they don't keep upgrading, not only will the market become saturated but the other chip manufacturers – AMD, Cyrix, and NexGen – will be able to catch up.

This is not a new problem for Intel; processing capabilities have always led the software applications. For example, although 32-bit processing has been a technological reality since 1985, Microsoft's first 32-bit processing system – Windows NT – didn't appear until

13

1993.[5] Intel has always been on the lookout for applications requiring massive processing capabilities.

One of the most CPU-intensive applications is video. Even the Pentium chip will not handle full-screen, 24-frames-per-second output. But the next-generation chip, the Pentium Pro, will. What Intel wants, therefore, is a cheap and widely used video application. To that end, it has invested over $100 million in ProShare, a videoconferencing system that sits atop a desktop computer.[6] ProShare is an ideal complement to Intel's chips.

But Intel faced the same problem that makers of fax machines faced a decade earlier: what's the point of having a desktop videoconference system if there's no one to call? Fax machines only took off in 1986, when their price came down to under $500. How could Intel establish a market presence for ProShare and get the price down without shelling out another $100 million? Intel's strategy was to look for other companies interested in helping out.

The phone companies proved to be one natural ally. ProShare complements their business because it receives and transmits more data than ordinary phone lines can handle. To work effectively, ProShare requires an Integrated Services Digital Network line, or ISDN line, for short.[7] These lines have three channels for transmission – two for data and one for voice – each with nearly five times the capability of ordinary twisted copper. The phone companies have the capability of supplying ISDN lines, but there's been little demand so far. If people buy ProShare, they'll buy ISDN lines, too.

So Intel doesn't have to pay for ProShare all by itself. Just as phone companies heavily subsidize the purchase of a new cellular phone in order to attract new cellular subscribers, some are subsidizing ProShare to encourage people to buy ISDN lines. They are offering ProShare to their customers for $999, less than half its list price of $1,999.[8]

In another move to create momentum for ProShare, Intel reached an agreement with Compaq, the leading PC maker, under which Compaq will include ProShare in all of its business PCs. This integration brings down the cost of ProShare for Compaq buyers to between $700 and $800 and gives ProShare's market presence a further boost.

All the players – Intel, phone companies, and Compaq – recognize their complementary relationship. Intel wants to increase the demand for processing capability; phone companies want to increase the demand for large amounts of data transmission; Compaq wants its business PCs to stand out from the competition. These objectives come together perfectly with personal videoconferencing.

2 The Value Net

We're now in a position to better answer our first question: if business is a game, who are the players and what are their roles? Customers, suppliers, and competitors. And one more category: people who provide complements. There's no word for people who provide complements, so we're going to propose one: *complementor*. This is the natural counterpart to the term 'competitor.' The fact that we had to coin a new word is proof that the vital role of complements has been largely overlooked in business strategy.

In the rest of this chapter, we're going to present a complete picture of the game of business. We'll explore the roles of all four types of players – customers, suppliers, competitors, and complementors – and the interdependencies among them. We'll see how the same player can have multiple roles. We'll define exactly what we mean by our new term 'complementor': it will even prove useful to give a definition of the familiar term 'competitor.'

It's a back-to-basics exercise. Focusing on one type of player or one type of relationship tends to produce blind spots. Taking in the wider picture reveals many new strategic opportunities.

To get things started, we introduce a schematic map to help you visualize the whole game. This map, the Value Net, represents all the players and the interdependencies among them. As we proceed, you might start thinking about how you'd draw a Value Net for your business. You'll see the Value Net we drew for our own business later on in this chapter.

Along the vertical dimension of the Value Net are the company's *customers* and *suppliers*. Resources such as raw materials and labor flow from the suppliers to the company, and products and services flow from the company to its customers.[9] Money flows in the reverse

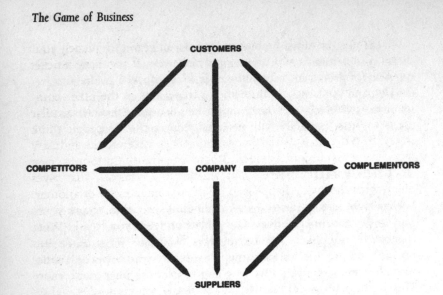

direction, from customers to the company and from the company to suppliers.

Along the horizontal dimension are the company's competitors and complementors. We've already seen many examples of complementors. Here's a definition of the term:

A player is your complementor if customers value your product more when they have the other player's product than when they have your product alone.

Thus, Oscar Mayer and Colman's are complementors. People value hot dogs more when they have mustard than when they don't. And vice versa. The way to identify complementors is to put yourself in your customers' shoes and ask yourself: what else might my customers buy that would make my product more valuable to them?

Competitors are the reverse case:

A player is your competitor if customers value your product less when they have the other player's product than when they have your product alone.

Coca-Cola and Pepsi-Cola are a classic example of competitors. So are American Airlines and Delta Air Lines. If you've just had a

Coke, you value a Pepsi a lot less than if you've yet to quench your thirst; Coke doesn't add life to Pepsi. Likewise, if you have a ticket on Delta, American is something a little less special in the air.

The traditional approach defined competitors as the other companies in your industry – those companies that make products similar to yours in a manufacturing or engineering sense. As people think more in terms of solving their customers' problems, the industry perspective is becoming increasingly irrelevant. Customers care about the end result, not about whether the company that gives them what they want happens to belong to one industry or another.

The right way to identify your competitors is, again, to put yourself in the customers' shoes. Our definition leads you to ask: What else might my customers buy that would make my product less valuable to them? How else might customers get their needs satisfied? These questions will lead to a much longer, and more insightful, list of competitors. Thus Intel and American may end up as competitors as videoconferencing takes off and becomes a substitute for business trips.

As Microsoft and Citibank each work to solve the problem of how people will transact in the future – whether it be E-money, smart cards, on-line transfers, or something else – they might end up being competitors. This is despite the fact that they come from different 'industries' as traditionally defined – software and banking.

Phone companies and cable television companies are both working to solve the problem of how people will communicate with each other and access information in the future. Again, different industries – telecommunications and cable television – but increasingly one market. Today European banks are selling insurance, and European insurance companies are selling tax-advantaged savings vehicles. It's no longer the banking industry or the insurance industry – it's one marketplace for financial services.

So far we've been stepping into customers' shoes to identify who complements you and who competes with you in attracting customers' dollars. But that's only half the game.

The supplier side

The top half of the Value Net deals with customers, the bottom half deals with suppliers. And, just as with customers, there are two sides to the game with suppliers. Other players can complement you or compete with you in attracting suppliers' resources. Here are the definitions:

**A player is your complementor if it's more attractive for a supplier to provide resources to you when it's also supplying the other player than when it's supplying you alone.
A player is your competitor if it's less attractive for a supplier to provide resources to you when it's also supplying the other player than when it's supplying you alone.**

Competition for suppliers often crosses industry boundaries. Capital providers are suppliers, and the competition to attract their funds takes place in a global market. Employees, too, are suppliers. People don't usually look at it that way, but follow the dollar: the company pays employees to provide a valuable resource – namely, their expertise, labor, and time. Competition for employees crosses industry boundaries. For example, companies from very different industries compete for the supply of newly minted MBAs.

Many companies are both competitors and complementors with respect to their suppliers. Compaq and Dell, for example, compete for the limited supply of Intel's latest chip. But the two companies are complementors as well as competitors with respect to Intel. Between development costs and building a new fabrication plant, Intel will spend well over a billion dollars to develop the next-generation chip. Intel will be able to spread that cost among Compaq, Dell, and all the other hardware makers, which means that each one of them will pay less to have Intel inside.

American and Delta compete for landing slots and gates. But although they are competitors for airport facilities, they are complementors with respect to Boeing, a key supplier. When American and Delta decide to commission the next-generation airplane, it's much cheaper for Boeing to design a new plane for both airlines together than to design a new plane for each of them separately.

Most of the development costs can be shared, and the greater demand helps Boeing move down the learning curve faster, too.

The same principle applies to fighter planes, although the US Congress may have discovered this a bit too late. The F-22 fighter jet complements and is complemented by other defense programs that share common development activity, such as avionics and navigation. Kill one of these supply-side complementors and you may shoot down the F-22 without meaning to. William Anders, former chairman and CEO of General Dynamics Corporation, explains the problem:

The F-22 is being recognized as one of the most successful, best managed next-generation weapons system development programs currently under way. However, as demand continues to fall in other defense programs served by the F-22 team, a portion of the fixed and overhead costs formerly supported by those programs automatically shifts over to the F-22. The danger is that this model program could ultimately become unaffordable because of the growing overhead and fixed cost burdens.[10]

In cutting back defense programs that it decides it can do without, Congress inadvertently endangers programs that it wants to keep. With complements, it's sometimes all or nothing. There may be no halfway.

As we continue moving into the information economy, supply-side complementarities will become increasingly the norm. There's a big up-front investment in learning to make something – whether computer chips or airplanes – and then variable costs are relatively modest. There's huge leverage. The more people that want a knowledge-based product, the easier it is to provide.

In the case of computer software or drugs, essentially *all* the costs are up-front; then it's gravy. For Microsoft, all the real cost comes in the intellectual step of writing the computer code for a new program. Copying the disks costs only pennies. So the bigger the market, the more the development costs can be spread out. The mass-market program is better and cheaper than what any one person could commission on his own. That's the nature of markets for knowledge-based products.

Recognizing symmetries

The Value Net reveals two fundamental symmetries in the game of business. On the vertical dimension, customers and suppliers play symmetric roles. They are equal partners in creating value. But people don't always recognize this symmetry. While the concept of listening to the customer has become a commonplace, the same isn't true when it comes to suppliers. We've all heard people tell their suppliers: 'You've got the specs. You don't need to know what the product's for. Just get it to me on time at the lowest price.' Imagine talking to customers that way! Only recently have people begun to recognize that working with suppliers is just as valuable as listening to the customer.

Supplier relations are just as important as customer relations. In one labor negotiation, we heard the head of human resources exclaim: 'I have to get my employees to understand that the customer comes first.' Seeing the Value Net helped change his mind and made for much more productive discussions. In the end, everyone recognized a common goal – to create the biggest pie. To do that, you can't put anyone first. If a customer wants something special, such as rush delivery, but isn't willing to pay enough to compensate workers for a lost weekend with their families, then satisfying this order would not create value – in fact, it would destroy value. The customer isn't always right. Employees have rights, too.

On the horizontal dimension, there's another symmetry. Go back to the definitions of complementor and competitor. You'll see that the only difference between them is that where it says 'more' in the definition of complementor, it says 'less' in the definition of competitor. At the conceptual level, complementors are just the mirror image of competitors. That's not to say that people are equally good at seeing the mirror image. Just as people have been playing catch-up when it comes to thinking about suppliers, there's a lot more work to be done in recognizing and benefiting from complementor relationships.

SYMMETRIES OF THE VALUE NET

Customers and suppliers play symmetric roles.
Competitors and complementors play mirror-image roles.

It's easy to focus on only one part of your business, and miss others. The Value Net is designed to counter this bias. It depicts all four types of players you interact with, and it emphasizes the symmetries between them – the symmetries between customers and suppliers, and between competitors and complementors.

3 Surfing the Net

To understand the game you're in, start by going around your Value Net. This approach applies to any organization – private, public, or nonprofit. As an example, we'll take you on a tour of the Value Net we know best – namely, our own.

Drawing a Value Net for a university gave us a better understanding of some of the issues facing our home institutions.[11] And, in our consulting work, we've found that the best way to begin assignments is by helping clients draw their own Value Nets. This exercise is an essential input into the process of generating new strategies, as we'll see later in the book when we lay out our PARTS model.

The university's customers

Who are the customers of a university? Students, primarily. Strangely, universities don't always treat their students as if they are the customers. Some people say that's the way it should be; after all, the faculty has know-how that students don't yet have. To the extent that's true, we say that makes students 'clients.' They're employing a professional service – like a doctor or lawyer – and should follow the guidance of the faculty. And, in return, the faculty should listen carefully when students express satisfaction or dissatisfaction with the service they get.

Universities have other customers. Parents are customers when they pay for their children's education. Companies are customers

CUSTOMERS

Students, Parents,
Federal Government,
State Government,
Companies, Donors

COMPETITORS

Other Colleges,
Freelancing Faculty,
Private Enterprise,
Hospitals, Museums

THE UNIVERSITY

COMPLEMENTORS

Other Colleges,
K–12 Education,
Computers, Housing,
Airlines, Hotels,
Cultural Activities,
Local Employers,
Copy Shops

SUPPLIERS

Faculty, Staff, Administrators,
Publishers (books, journals,
online services)

of business schools when they pay to send employees there, or when they pay schools to put on special programs for them. And the government is a customer when it pays for tuition scholarships. The government is a customer of a different sort when it commissions research from universities.

Another very important customer group is donors. Donors as customers? Yes. They seek fulfillment, prestige, or the opportunity to shape future generations in return for their contributions. Viewing donors as customers might give some universities pause for thought. Too often, a university fund-raising campaign starts with a list of priorities – a 'mission' – and tries to persuade donors to fund them. This is hardly listening to the customer. Like all customers, donors are free to take their 'business' elsewhere. Perhaps universities should pay more attention to what donors want. Asking donors what they would like to fund would build better relationships and likely raise more money, both now and in the future.

Of course, these different customer groups sometimes have competing views as to what type of education a university should provide. A university may not be able to listen to all of its customers at the same time.

The university's suppliers

A university's suppliers are primarily its employees: faculty, staff, and administrators. Since universities are in the business of disseminating information, they are also in the market for ideas. Thus, publishers of academic books and journals and providers of electronic information services (such as Lexis/Nexis and WestLaw) are suppliers to universities, too.

The university's competitors

Universities have no shortage of competitors: admissions offices compete with one another for students; faculty across schools compete for government and foundation grant money. Universities even face competition from their own faculty. For example, business school professors often give their own executive-education programs to companies on a freelance basis. That makes them competitors to business schools in the executive-education market.[12] Meanwhile, college presidents, along with their development officers, compete for the checkbooks of potential donors. They compete not only with other universities but also with hospitals, museums, and other nonprofits.

On the supply side, universities compete with one another for employees – faculty and administrators, in particular. Sometimes they also have to compete with private enterprise for talent. For example, finance professors Myron Scholes from Stanford and the late Fisher Black from MIT, inventors of the Black-Scholes option pricing model, left academia for Wall Street.

Technology is likely to increase the competition among schools. As videoconferencing becomes better and cheaper, remote classes will grow in importance. The university with the best undergraduate biology course, say, will be able to offer that course to students everywhere. This, in turn, will make universities less reliant on any but the very best faculty.[13]

The university's complementors

Universities, though they compete with one another for students and faculty, are complementors in creating the market for higher education in the first place. High school students are more willing to invest in preparing for college, knowing that there are many schools to choose from. More college students pursue Ph.D.s, knowing that there are a number of schools that might hire them.

The list of complementors to a university is huge. Kindergarten, elementary, junior high, and high schools complement universities. The better a student's earlier education, the more he or she will benefit from a college education. By the same token, undergraduate schools in one university and graduate schools in another university complement one another. The better a student's undergraduate program, the more he or she will gain from a graduate program.

Other complements to a college education are computers and housing. That's why most universities help their students buy computers and locate off-campus housing. Since schools attract students from all over, perhaps they should also help their students buy air travel – another complement. Working as consultants to Sallie Mae, the largest student loan provider, we've been able to put this perspective into practice. Sallie Mae is now helping its students buy complements for less. It has negotiated special student discounts on Northwest flights, MCI phone calls, and some publishers' textbooks.

Local hotel accommodation is an important complement to business schools offering executive-education programs. Accommodations were a problem for Northwestern's Kellogg School of Management, since there were few high-quality hotel rooms in Evanston, Illinois. So Kellogg built its own executive-education quarters.

Cultural activities and restaurants make a university more attractive to students. In this respect, schools in New York and Boston, for example, have an edge over those in Palo Alto and Princeton. There are many, many other complementors to universities – 24-hour copy shops, coffee shops, pizza and ice cream parlors, and more. These businesses all make a point of locating close to university campuses.

Local employers complement universities. Thus, as dual-career couples become increasingly common, Harvard has an advantage over Yale because of all the other Boston-area businesses that offer employment opportunities as compared with the depressed New Haven economy. To overcome this disadvantage, Yale is going to have to work harder at helping spouses with job placement, or be more willing to hire spouses at the university.

No doubt there's more to say about the Value Net for a university. The larger point is that drawing the Value Net for your business is a valuable exercise. You may already know the business inside out. Drawing your Value Net requires you to understand your customers' and suppliers' perspectives: you'll be forced to know the business 'outside in,' too.

Multiple perspectives

So far we've been surfing the Value Net from only one vantage point. You put yourself in the middle and then look around to your customers, suppliers, competitors, and complementors. Of course, that's not the whole game. There are also your customers' customers, your suppliers' suppliers, your competitors' competitors, your complementors' complementors, and the list goes on. For example, recruiters who come to hire graduating students are a university's customers' customers.

You could try to draw an extended Value Net to represent these extended relationships, but it would quickly become a mess. The better way is to draw multiple nets. Draw a separate Value Net from each perspective: your customers', your suppliers', your competitors', and your complementors', and, perhaps, from perspectives even further removed. For example, drawing your customers' Value Net might help you find a way to increase the demand for whatever your customer sells. Helping your customer this way helps you, too.

4 Playing Multiple Roles

> **All the world's a stage,**
> **And all the men and women merely players:**
> **They have their exits and their entrances;**
> **And one man in his time plays many parts . . .**
> — Shakespeare, *As You Like It*

People play many parts in the game of business. That makes the game a lot more complicated. Sometimes you see someone occupying one role and forget to ask what other roles that person plays. Other times you can't seem to fit someone into any particular role at all and then discover that's because that person is playing two or more roles simultaneously. The Value Net enables you to sort through all this tangle.

We've already seen some examples of how players can occupy more than one role in the Value Net. From the perspective of American Airlines, Delta is both a competitor and a complementor. American and Delta compete for passengers, landing slots, and gates, but complement each other when commissioning Boeing to build a new plane.[14] For American, it would be a mistake to view Delta solely as a competitor or solely as a complementor – Delta plays both roles.

It's the norm for the same player to occupy multiple roles in the Value Net. Strategy experts Gary Hamel and C. K. Prahalad give an example in *Competing for the Future*: 'On any given day AT&T might find Motorola to be a supplier, buyer, competitor, or partner.'[15] It won't be too long before electric utilities use their lines to transmit voice and data along with electricity. At that point, they will become competitors to the phone companies. But that hasn't stopped Southern New England Telephone and Northeast Utilities from becoming complementors today: the two companies run phone and electric wires over a common set of poles, enabling both to save money.

In the nonprofit sector, the Museum of Modern Art (MOMA) and the Guggenheim Museum in New York compete for visitors, members, and curators as well as for paintings and funding. Still, it's not all competition. The option to visit several museums in a weekend helps bring people into New York. Thus, the Guggenheim is a complementor as well as a competitor to MOMA. Perhaps the museums should create a joint weekend pass, a common practice in many European cities.

There's more. The Guggenheim might borrow a painting from MOMA or lend MOMA a painting to create a special show. Then the Guggenheim becomes a customer and supplier as well as a competitor and complementor to MOMA.

The position in the Value Net merely represents a role someone plays, and the same player can have multiple roles. It's counterproductive to typecast someone as just a customer or just a supplier or just a competitor or just a complementor.

Jekyll and Hyde

People are so accustomed to viewing the business world in warlike terms that even when other players are both competitors and complementors, they tend to see them as only competitors and fight against them. They focus on the evil Mr Hyde and overlook the good Dr Jekyll.

In the early 1980s, when sales of videocassette recorders took off, the movie studios were convinced that people might not see a film at the movie theater if they knew they could rent or buy it in the future. Although the studios would make money from video rentals and sales, this business would so significantly eat into their big-screen profits that they would end up worse off than before. So studios priced their movies sufficiently high that rental stores could only afford to buy a few copies each. Almost no movies were sold directly to consumers.

The studios' concern over cannibalization had merit. Some people did indeed skip going to the movies and were content to wait for the video release. But there was a much more important complementarity effect. Movies that did well in the theaters whetted people's appetite to rent or buy the movie. Those who enjoyed the movie might themselves rent or buy it to see it again, or tell others who missed it on the big screen to be sure to see the video.

Now that the studios have caught on, they have begun offering videos for sale at prices below $20 rather than selling only to video rental stores for $69.95. As a result, the combined market for movie theaters, video rentals, and video sales is far greater than in the days before video. In 1980 the industry revenue from theatrical releases totaled $2.1 billion while home video brought in an additional $280 million. By 1995, theatrical releases were bringing in $4.9 billion; even better, home video rentals and sales totaled $7.3 billion.[16]

Just as the movie studios feared the home video market, traditional bookstores see electronic publishing and the Internet solely as competitors. Once again, they see only half the picture. What bookstores fail to recognize is that sales in one domain may stimulate demand in the other. According to McGraw-Hill CEO Joseph L. Dionne, 'In ten instances when we created an electronic version of the print edition . . . [demand for] the print version grew, too.'[17]

By helping the entire market grow, booksellers on the Internet, such as Amazon.com and BookZone, stimulate traditional book sales. Although sometimes people buy books from Amazon.com instead of a traditional bookstore, Amazon.com provides a place to buy books on the spur of the moment – at 2:00 A.M., for example. This extra sale helps enlarge the pie, but that's not all it does. Books are sold by word of mouth; one sale can create a chain reaction. If the Amazon.com customer likes the book and tells friends about it, they might then buy it at a traditional store. Or people may buy a book in the bookstore because their interest was piqued by an electronic book review on the Internet. And, ultimately, if the Internet helps sell more books, authors and publishers will produce more books, which is good not only for booksellers but also for customers.

In an article in *Publishers Weekly*, BookZone president Mary Westheimer responded to the cool reaction she received at the 1995 American Booksellers Association convention:

If, instead of fighting futilely, these threatened booksellers looked through the other end of the telescope, they might see that what they perceive as competition is actually a complement . . . Together, we can create an appetite that feeds our industry.[18] . . . If all of us – booksellers, publishers, distributors, authors – do a good job of selling, more people will buy more books. And if we all work together toward the goal, we and our customers, the readers, will be that much happier.[19]

While the traditional booksellers see only competition, Westheimer recognizes the elements of complementarity in the book business.

A dramatic example of mistaking Jekyll for Hyde is to be found in every office. When computers were first introduced, almost everyone thought they would eliminate 'paperwork.' Words and data stored as minute magnetic impulses seemed like the greatest competitor paper had ever faced. There was talk everywhere of 'paperless offices.' People even began to wax nostalgic about old-fashioned printed and written material. But the reality was very different. According to the *Wall Street Journal*, 'Despite rising cost of paper and increased use of computers, offices are projected to use 4.3 million tons of paper this year [1995], up from 2.9 million tons in 1989. Offices will consume 5.9 million tons by the year 2000.'[20] What computers really did was make paperwork easier to generate. To date, computers have complemented far more than they have competed with paper.

Even when they recognize a complement, some people turn it down. Citibank was the first bank to introduce the ATM, back in 1977. When other banks came along with their own ATMs, they wanted Citibank to join their networks. That would have made everyone's ATM cards more valuable. When banks are on a common network, each machine complements all the others. But Citibank refused to join. It didn't want to do anything that might help its competitors. It didn't want to help Dr Jekyll if that also meant helping Mr Hyde. That decision came at the expense of Citibank's own customers. Over time, the other bank networks became the national and international leaders, and Citibank customers were left out. The limited ATM access cost Citibank market share. Citibank eventually woke up. In 1991 it reversed course and joined the other networks.

Of course, a player can start out as Dr Jekyll and turn into Mr Hyde. Cable television originally complemented network television stations by extending their reach into towns with poor reception. However, over time, the cable companies started to broadcast more and more alternatives to the networks' programs: HBO, CNN, MTV, Nickelodeon, Nashville, Home Shopping Network, and many more. Even people with good reception started signing up for cable. As viewers turned to cable programming rather than network shows, cable turned into more competitor than complementor to the networks.

Why is it that people tend to see Mr Hyde and miss Dr Jekyll? It's the business-as-war mindset. Everything is competition, not complementarity. This perspective leads you to suppose that when customers buy someone else's product, they're less likely to buy yours. Or that when suppliers provide resources to someone else, they're less able to supply you. It's all competition.

Perhaps this mindset stems from the view that life is all about making trade-offs. No one can have it all. With only so much money, so much time, so many resources, people have to make choices. Customers and suppliers have to choose between you and the competition. It's 'either-or,' not both.

But that's not always true. The trade-off mindset fails to take account of complements. When a customer buys a complementary product, that makes him more likely to buy yours. It's both, not one or the other. Or when a supplier provides resources to a complementor, that makes it easier for him to supply you as well. Again, it's both, not one or the other.

To help recognize Dr Jekyll as well as Mr Hyde, remember to think complementors as well as competitors.

JEKYLL AND HYDE

There's a bias toward seeing every new player as a competitive threat.

But many players complement you as well as compete with you.

Look for complementary opportunities as well as competitive threats.

Making markets

The fact that players can be both competitor and complementor explains what otherwise appears to be strange behavior. At first blush, it seems quite peculiar that competing businesses often locate literally right next to one another: New York diamond merchants along 47th Street; art galleries in SoHo; antiquarian bookstores in London; movie cinemas in Westwood, Los Angeles; and car dealers located along a strip. In Brussels, antique stores are lined up all around the Place du Grand Sablon.

Shouldn't the antique stores spread out over Brussels so that each store could have its local market? There would then be less direct price competition, since customers would find it less convenient to compare prices. After all, Wal-Mart doesn't line up next to Kmart, and Pearle Vision doesn't shadow LensCrafters. Nor do coffee shops or dry cleaners generally congregate.

But that way of thinking sees the Brussels antique stores only as competitors. By locating near to one another, the antique stores become complementors, too. Instead of having to choose only one store to go to — possibly the wrong one — shoppers can go to the Place du Grand Sablon, browse, and make a more informed choice. And because it's a lot more convenient, people are more willing to set out to buy antiques in the first place. They can also be more confident that the merchandise will be high-quality, because a store with inferior products or inflated prices will have a much harder time staying in business if the superior competition is located right next door. Also, people are more willing to buy one store's table if they find the perfect chairs for it, and there's a good chance of finding those chairs in another store nearby. Making it easier to find chairs helps sell tables, and vice versa. By locating close together, antique stores, though competitors in dividing up the market, become complementors in creating the market in the first place.

In some cases, the bunching effect helps create a bigger market for suppliers as well as for customers. That's the case for the performing arts on and off Broadway, New York. On the customer side, all the different shows help bring people into the city, even though the shows compete for the same audiences on any given night. On the supply side, the bunching of performing arts creates a critical mass that makes it easier for all to attract suppliers. Chamber music can share the same stage as symphonies. Theater can share the stage with dance. Musicians who play symphonies can perform in operas and musicals. Costume designers for theater can work for opera and dance. Lighting designers can work across theater, music, opera, and dance. Directors have their choice of actors and actresses working off Broadway – or even in restaurants.[21]

Whether it be diamond merchants, art galleries, antiquarian bookstores, movie cinemas, car dealers, antique stores, or performing arts, bunching together creates complementarities that develop the market, even if there's sometimes more competition in dividing it up.

Toys 'R' Us seems to follow the opposite strategy, relying instead on destination shopping. Its stores typically are located in low-rent areas off highways, not next to other retailers. People go to Toys 'R' Us because it's specifically where they want to shop. Is Toys 'R' Us doing the right thing? We don't suggest that Toys 'R' Us locate next to another toy store, but perhaps having a McDonald's restaurant or a Discovery Zone (a supervised indoor kids' playground) inside the store would make going to Toys 'R' Us more appealing. After all, people on their way to the Discovery Zone for a birthday party would now pass through the toy store. How convenient. And why not have a Big Mac while you're there?

So far, it's all complements. But the problem for Toys 'R' Us is that parents who drop their kids off at the McDonald's or Discovery Zone while they're shopping will be less influenced by their children to buy toys on impulse. Thus, McDonald's and the Discovery Zone are complementors when it comes to getting people into the store, but perhaps competitors when it comes to what they buy. We don't know which effect dominates, but, with well over five hundred stores, Toys 'R' Us might benefit from running some experiments. In fact, Toys 'R' Us might be able to learn from its own operations abroad. In Japan, for example, Toys 'R' Us has joined forces with McDonald's and Blockbuster Video to create family malls.

PEACE AND WAR

Companies are
- complementors in making markets
- competitors in dividing up markets

A player you can't avoid

The ultimate example of a player occupying more than one position in the Value Net is government, both federal and state. Depending on the aspect of government you're looking at, it can appear in the role of customer, supplier, competitor, or complementor. It also has an important behind-the-scenes role.

When the government buys goods and services, it's being a customer like any other – only bigger. In its role as customer, the government commissions new roads, bridges, buildings, and prisons; buys medical care and education; procures vast amounts of military equipment. The government is also a supplier. Among other things, it sells oil and mineral rights, logging rights, and the rights to use radio spectrum.

When people pay taxes, they have less money to spend on other goods and services. In this way, the government 'competes' with private business for people's dollars. Admittedly, the 'competition' is rather one-sided, since paying taxes is obligatory. Likewise, when the government borrows money, it competes with companies looking to raise capital. State colleges compete with private colleges. The US Postal Service competes with Federal Express. As the country's biggest employer, the government also competes with any business looking to hire people.

Meanwhile, the government serves as a complementor to every business activity by providing basic infrastructure and civil order. Virtually every business depends on government for things like protection of life and property, a transportation network, civil courts, a stable currency, and so on. Without these things, people couldn't do business.

Along with its transactional roles as customer, supplier, competitor, and complementor, the government has the power to make laws and regulations that govern transactions among other players. We'll talk more about this behind-the-scenes role of the government as a rule maker in the Rules chapter.

5 Friend or Foe?

**Michael Corleone: Keep your friends close,
but your enemies closer.**
— The Godfather, Part II

In the game of business, who are your friends and who are your foes? Sounds like an easy question. You have three groups of friends and one group of foes, right? Customers, suppliers, and complementors are all on your side, while competitors clearly are not.

In fact, we know that can't be quite right. People understand, intuitively, that along the vertical dimension of the Value Net there is a mixture of cooperation and competition. It's cooperation when suppliers, companies, and customers come together to create value in the first place. But when the pie has to be divided up, customers press for lower prices, and suppliers want their slice, too. So it's competition when it comes time to dividing the pie. In the case studies throughout this book, you'll see the simultaneous elements of competition and cooperation at work.

What about the horizontal dimension? Who are your friends and foes here? You're pleased when a complementor enters the game, and, most of the time, you're happier if competitors stay out. So complementors are friends and competitors are foes? Yes. But, again, that's not the whole picture.

When a complementor enters the game, the pie grows. That's win-win. But then there's a tug-of-war with your complementor over who's going to be the main beneficiary. If your complementor gets less of the pie, that leaves more for you, and vice versa.

This tug-of-war between complementors is evident in the computer business. Since hardware makers complement Microsoft, Compaq's and Dell's entry into the IBM-compatible personal computer market benefited Microsoft. But Microsoft gains even more every time Compaq or Dell starts a price war. When the price of hardware falls, more people buy computers, which leads to more software sales. Microsoft wins. Even people who would have bought computers at the old, higher prices now have more money left over to buy software. Microsoft wins again. Complementors may be your friends, but you don't mind if they suffer a little. Their pain is your gain.

In fact, if your complementors are too comfortable, that may leave you little, if any, pie. For on-line services, a critical complement is local phone service. If phone calls are expensive, on-line services have to compensate with low prices, perhaps so low that they can't make money. That's a major reason why on-line services haven't taken off in Japan, where NTT dominates the telephone market and charges handsomely for local calls. By contrast, in most places in the US, local calls are unmetered, and this has helped fuel the explosive growth of America Online, CompuServe, and the multitude of Internet providers. Cheap complements are your friend.

What about competitors? Surely here, at least, the relationship is clear. It's survival of the fittest. It's war. Sometimes it is war. Later in the book, we'll see how Nintendo conquered all opposition to become a giant of the videogame business. We'll also see how NutraSweet fought a price war in Europe to establish a valuable precedent. But the idea that it's *always* war with competitors is overly simplistic. Often, the win-lose approach leads to a Pyrrhic victory. Win-lose becomes lose-lose. The classic example is cutting price in an attempt to steal market share. Competitors match your lower price, and the result is lower profits all around.

Another problem with waging war on competitors is that it can be very hard to kill them off. Often, you succeed only in wounding them, and the most dangerous animal is a wounded one. Now that you've lowered their profits, your competitors have less to lose and every reason to become more aggressive.

An alternative approach is to find win-win opportunities with competitors.

Win-win opportunities with competitors, really? People talk about cooperation inside the company, working in teams and sharing information. But stepping outside, it seems naive to think of letting competitors 'win.' It isn't. What matters is not whether others win – it's a fact of life that they sometimes will – but whether you win.

Although it's hard to get used to the idea, sometimes the best way to succeed is to let others do well, including your competitors. We've seen examples where companies regarded primarily as competitors are also complementors. Insofar as these companies succeed *as complementors*, they are clearly benefiting each other.

Thus, you may want to work together with your competitors in order to develop common complements. In the early days of the automobile, competing carmakers worked together to build roads. Today, competing hi-tech companies are regularly joining forces to build infrastructure and standards for the information economy. For example, to help make Java the next standard for networked computing, IBM, Sun, and Compaq, along with Cisco, Netscape, Oracle, and several others, came together in August 1996 to create a $100 million venture capital fund to promote Java technology.

There are also times when the best strategy is to let competitors succeed *as competitors*. In the remainder of this book, we'll see cases where a company's move to undercut competitors could easily undercut the company itself. We'll show how to achieve win-win outcomes by avoiding mutually destructive competition. We'll explore how giving away a bid often comes back to hurt you, and suggest a better way to compete. We'll see how loyalty programs help everyone avoid falling into the price-war trap. We'll examine how rules such as meet-the-competition clauses change the nature of competition. We'll look at how to influence perceptions in order to shape competitive responses for everyone's benefit. In short, we'll discover situations where it's worthwhile to let a competitor prosper. A prosperous competitor is often less dangerous than a desperate one.

Julius Caesar: Let me have men about me that are fat.
– Shakespeare, *Julius Caesar*

Fine, but it's still a bad idea to turn your back. We are fully aware that your competitors may be all too happy to eat your lunch if you let them. Our recipe for strategy isn't about giving your competitors a free lunch – your lunch. We don't propose that you simply act nice, hoping that others will reciprocate. All too often that's a lose-win recipe.[22] We have something else in mind: a smarter way to compete that doesn't rely on the goodwill of others.

Your relationship with competitors is prima facie competitive, or win-lose. You lose when they enter the game. But you don't have to lose as much if you recognize that, once competitors enter the game, you can have win-win interactions with them. It's not all war with competitors. It's war and peace.

The same is true in all four directions. Whether it be customer, supplier, complementor, or competitor, no one can be cast purely as friend or foe. There is a duality in every relationship – the simultaneous elements of win-win and win-lose. Peace and War.

FRIEND AND FOE

There are both win-win and win-lose elements in relationships with

- customers
- suppliers
- complementors
- competitors

At this point, we have a map (the Value Net) and a mindset (co-opetition) for thinking about the game of business. We've seen some examples of how companies have changed the game, such as Ford and Ford Motor Credit, and Intel and ProShare, and hinted at many more examples to come. But we don't yet have a systematic method for how to change the game. To develop a method, we turn to game theory. That's the subject of the next chapter.

3 Game Theory

Life is the game that must be played.
– Edwin Robinson, poet (1869–1935)

How much can you hope to get in a game? As we'll see, the answer doesn't depend just on the size of the pie to be divided, or notions of fairness. Nor does it depend just on how well you play. What you get depends on your power in the game as well as on the power of others who have competing claims on the pie. Power – yours and others' – is determined by the *structure* of the game. Game theory shows how to quantify this power.

Game theory began as a branch of applied mathematics. It could be called the science of strategy.[1] It analyzes situations in which people's fortunes are interdependent. Game theory provides a systematic way to develop strategies when one person's fate depends on what other people do.

Game theory sounds tailor-made for the analysis of business strategy. But, historically, there's been an obstacle preventing the world of business from embracing game theory. The problem is that academics and businesspeople speak two different languages: equations versus experience. Many businesspeople have heard of game theory and suspect that it's a potentially powerful tool. But all the mathematics can be baffling and stops people from connecting the theory to practice. At the same time, game theorists are often unfamiliar with business practice, and some of their theories don't capture reality. Our experiences in teaching, research, and consulting suggest that communication between the worlds of game theory and business practice is both possible and valuable. This book brings theory and practice together.

In this chapter, we explain the fundamental ideas of game theory. In the rest of the book, we'll focus on the application of game theory to business strategy. Here, we're laying the foundations, trying to develop a new way of thinking. To do that, we use some deliberately simple and stylized games designed to illustrate the basic concepts of game theory. We've left out the mathematics, but the reasoning still requires close attention. If you, the reader, invest some time in this chapter, we promise you a big payoff in the chapters that follow, in which we apply these concepts to analyze and develop a wide variety of business strategies.

It's All in the Cards To see how game theory works, we'll start with a deceptively simple game. It's a slow day at Harvard, and Adam and twenty-six of his MBA students are playing a card game. Adam keeps the twenty-six black cards and distributes one red card to each of the students. The dean is feeling generous and agrees to put up $2,600 in prize money. He offers to pay $100 to anyone — either Adam or a student — who turns in a pair of cards, one black and one red.

That's the game. It's a free-form negotiation between Adam and the students. The only stipulation is that the students can't get together and bargain as a group with Adam. They have to bargain on an individual basis. Where would you expect the negotiations to end up?

Imagine that you are one of the students, and Adam offers you $20 for your red card. Would you take it?

We've played this game many times — with students, managers, executives, marketers, labor negotiators, and lawyers. People's first reactions are almost always the same: Adam is in the stronger position. From the students' perspective, Adam is literally holding all the cards. If they want to make a deal, they have to go to Adam. He has a monopoly on the black cards. Thus, Adam should do extremely well in the bargaining.

Are you ready now to take Adam's offer of $20?

Not so fast. Your position is more powerful than it may at first appear, so go ahead and turn down Adam's $20 offer. Perhaps you counter with a demand of $90. Don't worry if Adam rejects your counteroffer. Sit tight. Even if you and Adam can't agree on a deal right now, the game's not over.

Adam negotiates deals with each of the other twenty-five students. What happens next? Adam still has one black card left, and there is still one red card out there. It belongs to you. To make that last deal, Adam needs you just as much as you need Adam. With you and Adam now in completely symmetric positions, neither of you has an edge in this one-on-one bargaining. A 50/50 split is the most likely outcome.

By waiting, you can get $50 for your red card. Since the eventual deal will be 50/50, Adam and you might as well agree to a 50/50 deal up front. And since any student can play your strategy, the outcome is likely to be 50/50 all around. The game really comes down to twenty-six separate bilateral negotiations. To accomplish each deal, Adam needs the student just as much as the student needs Adam.

Barry then decides to try the same game back in New Haven. But as he stands in front of the class, it becomes apparent that Barry is not playing with a full deck: he's missing three of the black cards. An unfortunate accident, it seems. Barry plays the game with twenty-three black cards and distributes the twenty-six red cards to his students. As before, a black card and a red card together are worth $100. Where will the bargaining between Barry and his students end up? With a smaller pie to go around, will Barry and his students end up worse off than Adam and his students?

Once again, put yourself in the class. Barry offers you $20 for your red card. Would you take it, or would you hold out for more?

If you try your previous bargaining strategy, you'll be in for a surprise. This time, holding out is a bad idea. Because Adam had twenty-six cards, he needed all twenty-six students in order to make all the matches. If you turned down Adam's initial offer, you could count on his coming back. But with only twenty-three cards, Barry is playing a game of musical chairs, and three students will be left out. Should you turn down the $20 and counter with $90, Barry might walk away and never come back to you. You'd end up with a red card and no cash.

What holds for you holds for everyone else. Any student who doesn't agree to Barry's terms faces the prospect of being left out. So, one at a time, the students give in. Twenty-three 'lucky' students get $20 and three end up with nothing. If Barry offers you $20, take it.

Indeed, Barry could even propose a 90/10 split. There are three students who face the prospect of ending up with nothing. They'd happy to undercut those who hold out for $20. Anyone who ends up with $10 is still lucky. For Barry, 90 percent of $2,300 is a lot better than half of $2,600.

Losing three cards was no accident. Barry was a little Machiavellian. True, he made the pie a little smaller, but he understood very well how losing three cards would change the division of the pie. He knew that getting a sufficiently large slice would more than compensate for the reduction in the size of the pie.

Just a card game? No, a strategy employed by video game giant Nintendo, which, it just so happens, was originally a manufacturer of playing cards. In 1988–89 there was a shortage of Nintendo's video game cartridges. Nintendo chose to play Barry's version of the Card Game rather than Adam's, but with one big difference – it made a lot more money than either Adam or Barry. More on the story of Nintendo in the Added Values chapter.

Sacking the Cities The National Football League (NFL) scores big by playing Barry's version of the Card Game. By deliberately restricting the number of teams in the league, the NFL ensures that there are always more cities wanting football teams than there are teams. In 1988 the St Louis Cardinals moved to Phoenix, leaving St Louis with a football stadium but no team. In its bid to attract a replacement, St Louis made several overtures to teams. It didn't have much success; all it accomplished was to force cities with teams to match its generous offers. Finally, in 1995, St Louis persuaded the Rams to move from Anaheim to St Louis. Now Los Angeles has two empty stadiums, having previously lost the Raiders to Oakland. When the Baltimore Colts bucked Maryland and moved to Indianapolis, that left Baltimore eager to find a replacement. It took a new, publicly financed $200-million stadium and a $75-million up-front payment before the Cleveland Browns decided to make the greener pastures of Baltimore their new home. Where does that leave Cleveland? With an empty stadium.

More and more teams are now acting as free agents. With the lure of a $300-million state-of-the-art stadium and a $28-million relocation fee, the Houston Oilers want to move to Nashville. Then Houston will have – guess what – an empty stadium. Meanwhile, the Chicago Bears are thinking of moving to Gary, Indiana. The Tampa Bay Buccaneers may take a hike to Orlando.[2] The Seattle

Seahawks are considering flying the coop, as are the Phoenix Cardinals, yet again.

There are many cities chasing after not-so-many teams. That's why the teams do such a remarkably good job of negotiating stadium deals with municipalities. The teams have all the power; cities that want teams have comparatively little. As a result, even cities with teams don't see most of the benefits. A 1992 estimate put state and local government subsidies to team owners at $500 million annually. And that was when the competition for teams was only just beginning.

By playing Barry's version of the Card Game, the NFL has made money, but at a cost. As teams become less loyal to their hometowns, fans become less loyal to their teams. In the long run, that's bad for the NFL. We'll talk more about the pros and cons of undersupplying the market in the Added Values chapter.

The Card Game is a good story to bear in mind whenever you're trying to understand who has power in a game. We'll refer back to it several times in the chapters that follow.

1 Added Value

In the Card Game, we were able to argue our way through to who gets what. Game theory provides the general principle that explains who gets what in everything from the Card Game to the game of business to the game of life. The key to understanding who has power in any game is the concept of 'added value.'

Added value measures what each player brings to the game. Here's the formal definition: Take the size of the pie when you and everyone else are in the game; then see how big a pie the other players can create without you. The difference is your added value.

YOUR ADDED VALUE =

The size of the pie when you are in the game
minus

The size of the pie when you are out of the game

It's hard to get more from a game than your added value. Intuitively, what you can take away from a game is limited by what you bring, and what you bring is your added value. If you ask for more than you bring, what you've left for everyone else to divide is less than the pie they could create without you. Why should they agree to this? They could all do better by cutting a deal among themselves and leaving you out. So don't count on getting more than your added value.[3]

Let's go back to the Card Game and reexamine it through the lens of added value. First, look at the game with Adam and his twenty-six black cards. Without Adam and his cards, there's no game. Thus, Adam's added value equals the total value of the game, or $2,600. Each student has an added value of $100 because without the student's card, one less match can be made and, thus, $100 is lost. Therefore, the sum of the added values is $5,200 — Adam's $2,600 plus $100 from each of the twenty-six students. Given the symmetry of the game, it's likely that everyone will end up with half of his or her added value: Adam will buy the students' cards for $50 each or sell his for $50 each.

In Barry's version of the Card Game, the added values tell a different story. Because there are now only twenty-three black cards, the pie is smaller, $2,300. So is Barry's added value, also $2,300. But the more consequential effect of losing three cards is that each student now has zero added value. No student has any added value because three students are going to end up without a match. Therefore, no one student is essential to the game. The total value with twenty-six students is $2,300, and the total value with twenty-five students is the same, $2,300. Hence, each student's added value is zero. Barry is the only one with any claim to the pie. So, believe it or not, he's being generous when he offers the students $10 or $20 each.

There are some common errors that people make when they try to assess their added values. When we ask people to play the Card Game, we see these mistakes being made. The first error is to look at only half of the equation. People focus on the fact that, without Adam, they'd get nothing. They sense the weakness of their fallback option and ignore Adam's fallback option. They quickly agree to sell their card for a low price — often no more than $20 — and consider themselves lucky. But it's not enough to ask how much worse off you'd be without Adam. You also have to ask how much Adam stands to lose without you. Without you, he'd end up with none of that $100, either. So Adam's fallback option is really no better than yours. Consequently, Adam might pay any amount up to $100 for your card. That's your added value. Thinking in terms of added value helps you see the strength of your position.

To calculate your added value, ask: if I enter this game, what do I add? Instead of focusing on the minimum payoff you're willing to accept, be sure to consider how much the other players are willing to pay to have you in the game.

A second error is to confuse your individual added value with the larger added value of a group of people in the same position as you. We see this mistake when we switch to Barry's version of the Card Game. Students overplay their hand. They think they have some added value because without the students, Barry gets nothing. It's true that the added value of the students *as a group* is equal to the whole pie, namely $2,300. But that doesn't mean that the students are likely to get a large slice of the pie. That could happen only if all the students were to change the game by banding together and acting as a single player. Doing so would certainly be in their joint interest. It would be a strategy of changing the players, which is the subject of the Players chapter. But as long as they remain separate players, there's competition among the students to make a deal with Barry. And in that case the relevant number is the added value of any individual student, which is zero. That's why the students are lucky to get $20 in Barry's version of the Card Game.

Let's test-drive this last idea. The added value of cars is huge; we can hardly imagine life without them. But does that mean Ford has an equally huge added value? No. Take away Ford and you still have General Motors, Chrysler, Toyota, Nissan, and many other carmakers. What we *wouldn't* have is the Ford Explorer or the Ford Mustang. That's a much more limited sense in which Ford has added value — and a much more limited added value for Ford.

What is your added value?

It's always tempting to apply a new theory to one's everyday life. Thus, you can ask: what's my added value? But be aware that this is an uncomfortable question to entertain. You have to envisage what the world would look like without you. That's worse than reading your own obituary. It's like reading the newspaper one year later and seeing how the world got along without you.

In fact, some people work hard to keep their added value hidden. We all know people who refuse to stop work and take a vacation. They tell themselves – and anyone else who'll listen – that if they go away, the world will stop turning. If so, they really do have a large added value. More likely than not, the world would keep on spinning. Finding out that they're not irreplaceable after all is too much of a risk to take. So they keep working.

To look back and figure out what your added value was in some situation is the ultimate nihilistic experience. It's like imagining what the world would be like if you'd never existed at all. That's what Jimmy Stewart got to do in the movie *It's a Wonderful Life*.

In this 1946 story directed by Frank Capra, Jimmy Stewart plays George Bailey, banker, husband, and father. He has married his high school sweetheart (Donna Reed) and settled in his home-town of Bedford Falls to run his father's savings and loan business. George, who had dreamed of traveling the globe, feels trapped in his limited world. One day near Christmas, George's absentminded uncle takes the savings and loan's cash deposit to the bank but loses it before he reaches the teller. The evil Mr Potter (Lionel Barrymore), who owns the bank, finds the cash but doesn't return it to George. With the savings and loan effectively wiped out, George fears that Mr Potter will get his long-awaited opportunity to take over the business and, with it, the whole town. George falls into a deep depression and contemplates suicide. But then Clarence, an 'Angel Second Class' on a mission to earn his 'wings,' comes down from heaven to save George. Clarence shows George what the world would have been like if he had never been born. What George sees is a bleaker place: his life has had a high added value. His outlook on life reaffirmed, George proclaims that he wants to live, and a Merry Christmas to all.

We're going to do the George Bailey experiment many times in this book, both retrospectively and prospectively. We'll ask how the fortunes of the other players in the game would have been different had one of the players been absent. That's the experiment looking backward. We'll also do the experiment looking forward: how the future fortunes of the other players in the game would be affected if one of the players were to leave.

George Bailey had doubts about his added value, but Jimmy Stewart's added value was assured. Take him out of the picture, and *It's a Wonderful Life* wouldn't have been the same.

Established movie stars have enormous added value. That's a problem for the studios, which make the greatest returns on movies that don't have stars but still take off — *Rocky*; *Home Alone*; *Ace Ventura, Pet Detective*; and *Speed*. But the luck can't be repeated. The previously unknown actors and actresses — Sylvester Stallone, Macaulay Culkin, Jim Carrey, Sandra Bullock — are suddenly stars. The next time around, they have to be paid in line with their newfound added value.

When Macaulay Culkin was picked for *Home Alone*, he was just one of many aspiring child actors who could play the part. He had little added value and was happy to take the role for around $100,000. *Home Alone* went on to be the sixth highest grossing movie ever, and 20th Century-Fox grossed $286 million in the home market alone. There would surely be a sequel. But this time Macaulay had the added value.[4] To the moviegoing public, Macaulay Culkin was Kevin McCallister. The studio couldn't turn to another fresh face or even turn to another child star. So for *Home Alone 2: Lost in New York*, Macaulay got paid around $5 million, plus 5 percent of the domestic gross. The sequel grossed $174 million, and that added another $8.7 million to Macaulay's paycheck, helping him become the youngest of Hollywood's top-forty grossing actors.

Disney has a magic wand for creating unpaid stars. Its recent animated features — *The Little Mermaid, Beauty and the Beast, Aladdin, The Lion King,* and *Pocahontas* — have been some of the most profitable movies of all time. There are no human stars to pay, only animators. And the animators haven't been able to demand much of the pie because, individually, no animator has had much added value. That's because many animators work on each character, and no one person is irreplaceable. So Disney has kept the pie to itself. Today the animation game is in a state of flux. The added value of an animator is rising because Dreamworks SKG – the Steven Spielberg, Jeff Katzenberg, David Geffen outfit – is in the market alongside Disney to make animated features. On the other hand, Disney's smash success, Toy Story, replaces human animators with computer-generated imagery.

2 Rules

In the Card Game, there was no structure to the negotiations. Adam could make an offer to any of the students in any order, and the students could make any counteroffers they saw fit. If Adam and a student didn't reach an immediate agreement, the student could count on Adam's coming back later. There was no time limit on the negotiations.

Some negotiations are free-form, but others have rules. In business, you may have to give the same price to every customer. If so, that's a rule of the game. Or you may have a 'last-look' provision that gives you the option to keep your customer's business provided you match any competing bid. That could be another rule of the game. When supermarkets post prices on their goods, they're making a take-it-or-leave-it offer to customers. Another rule. In fact, this is the rule for most retailers. Nor is it just sellers who get to post prices; sometimes, buyers do, too. When employers go to buy someone's services, they often specify the terms and salary for the job, and there's very little room for further negotiation.

There are many rules governing negotiations in business. These rules come from custom, contracts, or law. Like added value, rules are an important source of power in games.

To see how rules can change the game, we return to the Card Game and add a simple rule: only Adam can make offers. As before, a black card and a red card together are worth $100. Adam has twenty-six black cards and each of the twenty-six students has one red card. But now it's a one-shot, take-it-or-leave-it negotiation – an 'ultimatum' game, if you like. As a student, you can either accept or reject Adam's offer. If you accept, the deal is done. If you reject, the game is over. You can't make a counteroffer; nor can Adam come back with a better offer. Either the deal is done or it's lost forever. Adam has only one chance with each student, and vice versa.

It's All in the Cards – a New Deal This time, imagine that you're playing Adam's role. How much would you offer a student for his or her card? Should you split the pie 50/50? Or can you do better than that? Alternatively, would you be lucky to do that well?

When we play this ultimatum game in the classroom, the results are remarkably consistent. We ask someone to play Adam's role. In the first negotiation, the person playing Adam typically offers $50, and this offer is accepted. The two parties split the $100 equally, and the first negotiation is over.

This result isn't surprising. The social custom of dividing the pie evenly is very strong. Anything else would be unfair. Offering less than $50 would be unfair to the student, and giving more than $50 would be unfair to Adam.

There's another consideration. If Adam were to offer less than $50, he might get turned down and end up with nothing. The student has the power to determine whether Adam will get any money at all. With the student in such a powerful position, it would be foolish for Adam to demand too much of the pie.

But that's only half the picture. The student will get nothing if he or she rejects Adam's offer. That seems to put the student in a weak position. So which is it? Is the student's position strong or weak? Is the power in the hands of the person who makes the offer or in the hands of the person who accepts or rejects it?

To find out, we call a time-out at the end of the first negotiation and debrief the players. To help the person playing Adam's role get the perspective of the other party, we allow him to confer with the student he's just bargained with before making an offer to the next student.

Typically, the student advises Adam to be much more aggressive. The student was prepared to accept much less than $50, perhaps as little as $5. So for the next negotiation, Adam offers a 90/10 split, with the $90 going to Adam. It's true that if this student says no, then Adam will end up with nothing, but so will the student. If the student focuses on the dollars, then the student will prefer $10 to nothing, and Adam will get his $90. A show of hands typically reveals that the vast majority of people — 95 percent is not uncommon — would take $10. A 95 percent hit rate on $90 is better than a guaranteed $50.

To analyze this game, put yourself in the student's shoes; recognize that the student is likely to accept the offer as long as he or she gets *some* money. The take-it-or-leave-it rule confers all the power to the person making the offer, none to the person accepting or rejecting it. As Adam, you can get a lot more than $50 if you play your cards right.

Of course, you shouldn't push your luck too far. If you were to offer only a penny, or even a dollar, the student might well turn you down out of pride or spite. You have to offer an amount that the other party considers better than nothing. Experience shows that with a $100 pie, one is extremely safe offering an 80/20 split, and even 90/10 is reasonable. But don't try 99/1.

In this ultimatum version of the Card Game, we were again able to argue our way through to who gets what. The general principle is that to every action there is a reaction. In physics, this is Newton's third law of motion. It's an equally true statement about games — but with an important difference. According to Newton's third law, the reaction is equal and opposite; in games, the reaction need not be equal or opposite. Reactions aren't programmed.

To anticipate other players' reactions to your actions, you have to put yourself in their shoes and imagine how they'll play the game. You look forward into the game and then reason backward to figure out which initial move will lead you where you want to end up. This principle applies to any game with a specified sequence of possible moves and countermoves.

This is exactly what we did in the ultimatum version of the Card Game. The rules were simple: Adam makes an offer, and the student either accepts or rejects. So, in this game, there was only one reaction to anticipate: either acceptance or rejection. Even so, the implications of the rule weren't entirely transparent. With more complicated rules, the implications can be harder to tease out. In the Rules chapter, we'll analyze the effects of many common rules in business — most-favored-customer clauses, meet-the-competition clauses, take-or-pay contracts, and more – to see how such rules can change the balance of power in a game.

3 Perceptions

Different people view the world differently. Just as the players' added values and the rules are important elements of a game, so are the players' perceptions. The way people perceive the game influences the moves they make.

Thus, any description of a game must include how people perceive the game – even how they believe other people perceive it, how they believe other people believe the game is perceived, and so on. There is no such thing as a game separate from the way the players perceive it.

Perceptions are particularly important in negotiations. Let's look at the classic negotiation problem of dividing up a pie.

Texas Shoot-Out When two partners set up a business or joint venture, they often include a rule specifying what to do if one of the partners wants to end the relationship. A common rule is the so-called Texas Shoot-Out. The dissatisfied partner states a price. The other partner must then either buy the first one out at that price or sell his partnership interest at that price. If you were initiating a shoot-out, what price would you pick?

Most people think that it's best to state a price at which you're equally happy being bought out or buying the other partner out. If you value the venture at $100 million, then you state a price of $50 million. You don't know what the other person will do, but this way you've guaranteed yourself half the pie.

Actually, you can do better than this. To see how, we take a short detour to examine the game 'I cut, you choose,' the kid's version of the Texas Shoot-Out. Two kids have to divide a pie – lemon meringue – and each wants the bigger slice. The classic rule is to have one cut the pie and the other choose which of the two slices to take.

The second child – the one who chooses which slice to take – must be at least as happy with his slice as with the other, since he had the choice. Anticipating this, the first child – the one who cuts the pie – realizes that he can't get more than half. If he divided the pie 60/40, then the second child would take the larger slice, leaving him with the smaller one. So the first child will do best to divide the pie evenly.

But this analysis is too simple. It presumes that the two kids view the pie the same way. Suppose, instead, that the second child prefers crust, and the first knows this. If that's the case, how should the first child divide the pie? He could still cut right down the middle, as shown on the left. (See next page.) But he can do better by cutting the pie as shown on the right. The top slice is bigger, but the bottom slice has just enough extra crust to entice the second to pick it, leaving the first with more than half the pie.

Let's apply this lesson to the Texas Shoot-Out. What matters isn't just what you think the venture is worth but also what you think the other partner thinks it's worth. The right strategy takes account of your perception of the other partner's perception of the pie.

Suppose you value the venture at $100 million, and you know your partner values it at $60 million. Then the right strategy is to figure out the price at which your partner would be equally happy to buy or sell, and to give him some incentive to make the choice that you want him to make. If you state a price of $50 million, you don't care if he buys or sells. But he would much rather sell to you at $50 million than pay that amount to buy something worth only $60 million to him. So you'd do better to state a price of $31 million. That way, he'd rather sell to you than buy you out – just. You'd get to buy his share for $31 million rather than $50 million. If you think it's better to give your partner more of a nudge in the right direction, then state a price of $35 million; you'll still come out ahead.

Is it reasonable to presume that you know your partner's valuation? Obviously, you can't know it exactly, but often you can have a pretty good idea. Remember, you *were* partners. The two of you have been working together, and so you have a good chance of understanding how your partner views the business. The two of you may even have debated how much the venture was worth. In a case where we helped a company through a Texas Shoot-Out, it was a disagreement over valuation that led to the breakup. Our

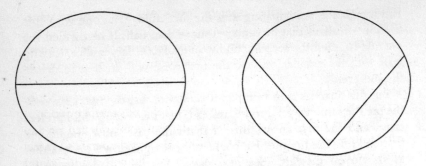

client wanted to make additional investments, but the other partner was less optimistic about the future of the enterprise and didn't want to put more money in. We felt confident in our assessment of the other partner's valuation and chose the shoot-out price accordingly.

What if you're in a Texas Shoot-Out and you don't have a good idea of how your partner perceives the venture? In this case, you really can't predict what he'll do. You could go back to the original approach: if you value the venture at $100 million, state a price of $50 million and thereby guarantee yourself $50 million regardless of whether your partner buys or sells. But there's another option to consider. Encourage your partner to shoot first. That way, you get the choice of whether to buy or sell. If your partner states a price below $50 million, buy; if he states a price above $50 million, sell. Either way, you net more than $50 million. If your partner happens to hit $50 million right on, you're no worse off than if you go first and state a price of $50 million. So in a case where you don't feel confident in your assessment of your partner's valuation, try to go second: there's no downside and a potential upside.

In the Texas Shoot-Out, the right strategy all depends on your perceptions. That's true in every game. Perceptions are always part of the picture. Sometimes they have a starring role.

Artistic Differences Midway through production of a certain big-budget action thriller, the director and the multimillion-dollar star had some 'artistic differences.' The director quit, and the studio scrambled to come up with a replacement. None of the obvious candidates was available, and with the production schedule slipping badly, the studio became desperate. It was prepared to spend whatever it took to hire a new director.

Meanwhile, the film's writer had proposed himself for the director's job. He'd never directed a major feature film before, but he had directed several episodes of a TV series. Most important, the star seemed to like him. The studio decided it was willing to take a chance on the writer as director. In fact, it didn't really see another option.

The writer didn't know that the production company had already tried to hire every director on the block. He was desperate to move into directing. He instructed his agent to get the best deal he could from the studio, but not to lose the deal over salary. If push came to shove, the writer was prepared to do the job for nothing.

The agent made the first move. He said his client would do the job for $300,000. The studio's lawyer had been instructed to see if he could hire the writer for $750,000 and was authorized to go up to $2 million. He was very pleasantly surprised by the agent's low figure but kept a poker face. Although he would have been delighted to agree to $300,000 immediately, he didn't want the agent to realize that he'd asked for too little and feel bad. He countered with $200,000, and after a little back-and-forth, the two parties settled at $250,000.

The agent was pleased to have landed the directing job for his client, and at a salary close to his initial offer. He had no idea how much money he'd left on the table. The studio's lawyer reported back that he'd saved the studio a half million, almost enough to justify his salary in one swoop. So everyone was happy?

Not quite. The studio head was happy that the movie was back on track. But when he was told how little the writer would be getting, he was horrified. He knew that once the star discovered what the studio was paying for a replacement director, the star would protest that he was being surrounded by second-rate talent. The studio head ordered that the directing fee be raised to $750,000. He also made clear that a certain lawyer wasn't going to be allowed to handle this kind of deal in the future. When the writer heard about his new fee, he was naturally thrilled; but he also decided his agent had been incompetent, and he fired him.

The story had a happy ending for the star, the studio, and the writer. Not so for the agent and the lawyer.

Where did the agent and lawyer go wrong? The agent failed to look at the game from the studio's perspective. He based his offer on his client's position and failed to consider how desperate the studio might be. He might have done better to let the studio make the first offer. The lawyer did some things right, at least in the small. He left the agent with the perception that he – the agent – had done the best he could in the negotiation. The lawyer's mistake was failing to recognize the larger game. He thought it was a game between the studio and the writer. He forgot about the real star of the show and how the star would feel about having a bargain basement director.

In the Tactics chapter, we'll have a lot more to say about the role of perceptions. We'll see how Rupert Murdoch's *New York Post* corrected a rival's misperception in order to prevent a price war. We'll return to the role of perceptions in negotiations. We'll also explain what the peacock's tail has got to do with perceptions, and the lesson it holds for business strategy.

The Artistic Differences story brings us to the last big piece in the picture of games we've gradually been assembling: the scope of the game. That's what the studio's lawyer missed.

4 Boundaries

So far we've introduced the concepts of added values, rules, and perceptions. There's one more element of a game: the scope of the game.

In principle, a game has no boundaries. There is really only one big game – extending across space, over time, down generations. But that's in principle. A game without boundaries is too complex to analyze. In practice, people draw boundaries in their minds to help them analyze the world. They create the fiction that there are many separate games.

Chess is a good example. No one can visualize it in its entirety, so people have created the fiction of the opening, middle game, and endgame. Business is no less complicated than chess, so business has its fictions, too. People often talk about a national economy, or an industry, as if it were the whole picture. Of course, everyone knows that's a fiction. In reality, the world's economies are highly interdependent – indeed, increasingly so. And, as we discussed in the Co-opetition chapter, industry boundaries are largely artificial.

Analyzing individual games in isolation is treacherous. You risk mistaking what is really only a part of the game for the whole. Every game is linked to other games: a game in one place affects games elsewhere, and a game today influences games tomorrow. The problem is that mental boundaries are not real boundaries.

Epson's entry into the laser-printer business is an illustration of what can go wrong when you get things right in the smaller game but miss the larger one.

Missing the Link In 1989 there were three types of desktop printers on the US market. Dot-matrix printers occupied the low end, laser printers the high end, with ink-jets in between. Dot-matrix printers accounted for about 80 percent of total unit sales of desktops, laser printers around 15 percent, with ink-jet taking the last 5 percent. Typical retail prices were $550 for a dot-matrix, $650 for an ink-jet, and $2,200 for a laser printer. At that time, Epson was the king of dot-matrix printers while Hewlett-Packard (HP) led in the laser and ink-jet segments.[5]

Looking at each of the three games – dot-matrix, ink-jet, and laser – in isolation suggested that Epson was in the wrong one. The laser segment had the highest prices and margins and was the fastest growing. So in August 1989 Epson launched a very competitively priced laser printer, the EPL-6000. It was a bit of a me-too product and lacked the Hewlett-Packard brand name. One week later, HP came out with its LaserJet IIP, priced significantly below the EPL-6000. Epson responded by reducing the price of the EPL-6000 and succeeded in building up to a 5 percent share of the laser printer business by December 1989.

Due to the intensifying price competition in the laser segment, other players, such as Toshiba, lowered the prices of their laser printers. Epson's gains stalled. The price competition also hurt HP's sales of its ink-jets. HP began promoting the ink-jets aggressively in order to counteract the narrowing price gap between the ink-jet and laser segments.

Epson then discovered that it was losing dot-matrix sales to the now comparably priced ink-jet machines. Prices had to come down in the dot-matrix segment, but there wasn't much room to go. Epson's core business was doubly squeezed.

What was Epson's mistake? It misunderstood the scope of the printer game. By treating the laser printer game as separate from the dot-matrix printer game, Epson failed to see that low-price entry into the laser segment could jeopardize its core business. Perhaps Epson assumed that high-end laser machines could never cannibalize sales of low-end dot-matrix printers. If so, it failed to think through the links from laser to ink-jet segment and from ink-jet to dot-matrix segment.

The Epson story shows how a move in one game can affect your fortunes in other games. The links between games can cause a cascade effect, and Epson didn't foresee the chain reaction it set off. Taken in the small, Epson's actions seemed reasonable, but looked at in the large, they weren't. And Epson failed to see the larger game. It didn't anticipate the competitors' reactions to its actions. If it had, it would have seen that it was much better off under the status quo.

In the Scope chapter, we'll return to the important subject of links between games.

5 Rationality and Irrationality

People often imagine that game theory requires all the players to be rational. Everyone is out to maximize profits. Everyone understands the game. There are no misperceptions. Feelings of pride, fairness, jealousy, spite, vengefulness, altruism, charity never arise. That's all very nice, but it's not the way the world is. So much for game theory.

In many ways, people are right, or *were* right. Granted, the simple textbooks present a view of 'rational man' that doesn't apply very well to the mixed-up, real world of business. But that's a problem with the textbooks. While early work in game theory didn't talk much about rationality or irrationality, current work does. The textbooks simply haven't caught up yet.

Early game theorists had good reason to spend little time worrying about irrationality. Game theory started out by analyzing zero-sum games, like poker and chess. In these games, failing to anticipate that the other player may make an irrational move doesn't get you into trouble. If he does something irrational, that's good news for you. Anything that makes him worse off must make you better off, since it's a zero-sum game.

But games in business are seldom zero-sum. That means you can succeed together or fail together. When another player can take you down with him, you care about his rationality. Think back to the Card Game. How Adam and a student divide the $100 is zero sum: if Adam gets more, the student gets less, and vice versa. But the fact that Adam and the student will both get nothing if they fail to reach an agreement makes this very much a non-zero-sum game. Either player, in hurting himself, hurts the other player, too. Each has to be concerned about the other's rationality.

What rationality is — and isn't

It's easy to get confused about just what 'rationality' means. Here's what it means to us: a person is rational if he does the best he can, given how he perceives the game (including his perceptions of perceptions) and how he evaluates the various possible outcomes of the game.

Two people can both be rational and yet perceive the game quite differently. One person may have better information than the other. But if the second doesn't know what the first knows, he's not being irrational in seeing things differently. The difference in information naturally leads to a difference in perceptions, even to mis-perceptions. People can guess wrong and still be rational. They're doing the best they can given what they know.

Likewise, two people can both be rational and yet evaluate the same outcome quite differently. People don't just look at the dollars. They're motivated by many things — pride, fairness, jealousy, spite, vengefulness, altruism, and charity are just a few of the possibilities. We saw this in the ultimatum version of the Card Game. A person need not be irrational to reject a very small offer, and if you're the one making the offer, you'd better recognize that. Throwing up your hands and asking, 'Why didn't he accept one cent?' doesn't make any sense.

People are very quick to deem others irrational when they see them doing 'crazy' things. In a case we encountered, senior management was ready to fire an 'irrational' salesman. He was so single-minded in going after volume that he cut prices to the point of destroying profits. He was a one-man price war.

But the salesman wasn't irrational. He understood all too well what determined his bonus. While, in theory, he was compensated on both sales volume and profit margins, he knew that when push came to shove, keeping the factory at capacity was what really mattered. In practice, his bonus depended on hitting and exceeding sales targets more than on maintaining profit margins. Instead of firing him, management came to see his perspective. The bonus compensation system was changed, and the salesman became a whole new person.

Simply dismissing someone as irrational closes the mind. Much better is to work harder at seeing the world as the other person sees it. This is a mind-expanding exercise. Trying to understand what motivates the other person, what drives him, can help you anticipate what he's going to do in the future or how he's going to respond to something you do.

In sum: the fact that other people view the world differently does not make them irrational. In fact, if you try to impose your rationality on others, who's the one who is really being irrational?

To us, the issue of whether people are rational or irrational is largely beside the point. More important is remembering to look at a game from multiple perspectives – your own and that of every other player. This simple-sounding idea is possibly the most profound insight of game theory.

Allocentrism

When I am getting ready to reason with a man I spend one-third of my time thinking about myself and what I am going to say, and two-thirds thinking about him and what he is going to say.
— Abraham Lincoln[6]

Many people view games egocentrically; they focus on their own position. The insight of game theory is the importance of focusing on others — namely, allocentrism.[7] This principle underlies everything we've said about added values, rules, and perceptions. To assess your added value, you have to put yourself in the other players' shoes and ask what you bring to them. To understand how a rule affects the play of a game, you have to put yourself in the other players' shoes to anticipate how they'll react to your move. To take account of differing perceptions, you have to put yourself in the other players' shoes and see how they look at the game.

The underlying principle is the same: you have to put yourself in the other players' shoes. You have to be allocentric. This doesn't mean you can ignore your own position. The skill lies in putting the two vantage points together: in understanding both the egocentric and the allocentric perspectives.

Putting yourself in other players' shoes does *not* mean: how would you analyze the game from their perspectives? It means: how would *they* analyze the game from their perspectives? It means putting yourself in their heads as much as in their shoes. It means adopting their views of the world. As part of that exercise, you also need to imagine how they perceive your view of the world. How will they put themselves in your head? Or rather, how do you think they will? And that's not the end. You even need to imagine how other players imagine you perceive their view of the world. How do they think you will put yourself in their heads? Or rather, how do you think they think you will? And so on. None of this is easy.

ALLOCENTRISM

Added Value: Put yourself in the shoes of other players to assess how valuable you are to them.
Rules: Put yourself in the shoes of other players to anticipate reactions to your actions.
Perceptions: Put yourself in the shoes of other players to understand how they see the game.

When you put yourself in other people's shoes, you'll find that they come in all different sizes. They certainly won't all fit your view, and allowing for these differences isn't easy. It can be uncomfortable giving credence to someone else's view of the world. There's a natural bias to impute your own views to other people. In *Getting to Yes*, leading negotiation experts Roger Fisher and William Ury offer some advice on how to overcome this bias:

The ability to see the situation as the other side sees it, as difficult as it may be, is one of the most important skills a negotiator can possess. It is not enough to know that they see things differently. If you want to influence them, you also need to understand empathetically the power of their point of view and to feel the emotional force with which they believe it. It is not enough to study them like beetles under a microscope; you need to know what it feels like to be a beetle. To accomplish this task you should be prepared to withhold judgement for a while as you 'try on' their views. They may well believe that their views are 'right' as strongly as you believe that yours are. You may see the glass as half full of cool water. Your spouse may see a dirty, half-empty glass about to cause a ring on the mahogany finish.[8]

There's an even more fundamental challenge when you try to put yourself in someone else's shoes. You know too much. It's like trying to play yourself in chess. You know what your own strategy is, but now you have to pretend that you don't in order to step into the other player's shoes. It's almost impossible to pretend that you don't know what you know.

The same issue arises when you try to figure out how someone will perceive or misperceive your perceptions of the world. In doing this, you're burdened by the fact that you know your own perceptions. Once again, how can you pretend that you don't know what you know?

One solution for how to step into the shoes of another player is to have someone else assist you. Instead of trying to do it yourself, ask a colleague to role-play by stepping into that player's shoes. Play out the game, see what you each do, and then debrief each other on what each perspective was like. What were the perceptions and misperceptions? Switch positions with your colleague and play the game again.

There's much to be gained from doing this exercise more formally. A company sets up two teams: one team plays out the company strategy, and the second plays the role of a competitor. The second team is given no advance information about the company's proposed strategy. It sees the strategy only as it unfolds and then must react to it. Quite often, the response is not what had been anticipated. Going through this exercise with clients, we've been able to help them avoid surprises in the real game.

Most of the time, putting yourself in other people's shoes helps get you where you want to go. Most of the time, but not always.

Crazy Driver There is a now somewhat infamous – yet true – story about one of your authors and a late-night taxi ride in Jerusalem. Some time ago, Barry and a colleague, John Geanakoplos, got into an Israeli taxi and gave the driver directions to their hotel. The driver headed off but didn't turn on his meter. When they asked him about this, he explained that he loved Americans and promised them a special fare. Special, huh?

As Barry and John sat in the backseat, they did their quick-and-dirty game theory analysis. They realized that if they bargained hard now and negotiations broke down, they might have to find another taxi, not an easy task. But, they reasoned, their position would be much stronger if they waited until they arrived at the hotel. Then the driver should be willing to take whatever they offered.

They arrived. The driver demanded 2,500 Israeli shekels ($2.75). Who knew if this was fair? But people generally bargain in Israel, so they counteroffered 2,200 shekels. The driver was outraged. Again, he demanded 2,500 shekels, and again they refused. Before negotiations could continue, he locked all the doors automatically and retraced the route at breakneck speed, ignoring traffic lights and pedestrians. Were they being kidnaped to Beirut? No. He took them back to where they'd started and ungraciously kicked them out of his cab, yelling, 'See how far your 2,200 shekels will get you now.'

When Barry and John found another cab, the driver turned on his meter, and 2,200 shekels later they were home. Certainly the extra time was not worth the 300 shekels. But, in the end, the trip was worthwhile. After all, it's a great story.

What went wrong? Perhaps it was a case of wounded pride. In retrospect, Barry and John might have paid more attention to the fact that the driver's girlfriend was sitting next to him in the front seat. Or perhaps the driver was simply crazy. Either way, Barry and John didn't push their game theory analysis quite far enough: next time, they'll get out of the taxi before discussing price.

6 The Elements of a Game

We've now introduced all the building blocks of game theory. Is that it? Well, yes and no. The concepts of game theory are simple, but deceptively so. Just knowing what these concepts are isn't enough. The trick is to apply the concepts creatively to a wide variety of real-world situations. To do this, we will need to use them to analyze much more complicated games than ones we've discussed in this chapter. The real power of game theory comes from taking this next step.

But first, it seems a good idea to recap briefly.

In the opening chapter, we pointed out how business isn't exclusively war; nor is it exclusively peace. We then described how game theory helps us get beyond those oversimplified outlooks. This gave us a new mindset and a jumping-off point for the material to follow.

The Co-opetition chapter identified the first and most fundamental element of the game of business: the Players. We used the Value Net to describe the cast of players and to diagram their relationships to each other. The complete list of players consisted of customers, suppliers, competitors, and complementors. We saw how useful it is for any business to identify all the players and their relationships, employing the Value Net.

This present chapter introduced the concept of *Added values*. These measure what each player contributes to the game by joining it. Added values can sound abstract, but they don't stay abstract for long. The added values determine who has power in a game and who will get the big payoffs.

After added values, we went on to discuss **Rules**. These structure the way the game is played. In business, there is no universal set of rules; they can come from custom, contracts, or law. Sometimes the most important rules are the ones taken almost for granted. We demonstrated how putting in a rule can produce a big difference in the way a game is played out.

Next we talked about perceptions. We showed how greatly any game is affected by the different ways different people perceive the situation. These differing perceptions are not just some subtle influence on the way the game is played. They are a fundamental part of the game itself. Equally important are perceptions of perceptions, perceptions of perceptions of perceptions, and so on. By altering players' perceptions, you can alter the moves they make. The devices used to shape perceptions are what we call *Tactics*.

Our discussion of perceptions led us into a discussion of the boundaries, or *Scope*, of the game. The issue here is the limits people implicitly place on games when they define them. Although people often analyze games in isolation, each game is invariably linked to others. To understand what's going on, you need to be sure to consider these links.

These, then, are the five elements of any game: *Players*, *Added values*, *Rules*, *Tactics*, and *Scope*. Putting them together gives us PARTS. It's important to recognize they are all components of a single whole. Sometimes the various elements will seem to overlap, because they depend so closely on each other. But we still need to look at each component individually to be sure none of them is being neglected. PARTS is the way to understand what's going on in any game.

Part 2

The PARTS
of Strategy

How to Change the Game

Philosophers have only interpreted the world.
The point, however, is to change it.
— Karl Marx

We don't mind being a little revolutionary in thinking that success comes from playing the right game. The biggest opportunities — and the biggest profits — don't come from playing the game differently. They come from changing the game itself. If you're playing the wrong game, you need to change it. Even if it's a good game, think about creating a better one. Changing the game is the essence of business strategy.

Marx had a point. The action comes from changing the game.

When people talk about changing the game, they often say: 'You have to change the rules.' That's certainly one way to change the game. But it's only one of several.

In fact, *every* element of the game is also a lever for changing it. To change a game, you need to change one or more of its elements. This means that each of the five elements — *Players*, *Added values*, *Rules*, *Tactics*, and *Scope* — gives you a way to transform an existing game into an entirely new one. Change one of the PARTS, and you change the whole.

To help you gain a deeper understanding of how to change the game, we devote a chapter to each of the five ways of doing so. We'll look first at changing the cast of players: how to do it and the consequences of doing it. Then, in the following four chapters, we'll go on to examine what's involved in changing the other four elements of a game. As you master each strategic lever, you'll become better equipped to change the game.

As long as you search for new strategies in a hit-or-miss fashion, you may well miss the best opportunities for your business. PARTS leaves nothing out. Go through all of PARTS in a deliberate way, and you can be confident of spotting *all* the possibilities. The strategies PARTS suggests may not always be new, although frequently they are. What's new is being able to generate strategies systematically.

PARTS does more than exhort you to 'think your way out of the box.' It provides the *tools* for finding your way out of the box.

4 Players

Cui bono?
Who stands to gain?
   – Cicero

Do you want to be a player? It's the obvious question to ask yourself when you consider entering a game. But the answer is rarely obvious. It's easy to misjudge what it would really be like to be in the game.

The reason is that anytime you enter a game, you change it. You don't have any choice in the matter. It's a new game because you've joined the cast of players. People often miss this effect. They fail to think through how their coming into a game will change it. They think that what they see is what they're going to get.

Not so. The game after you've entered it isn't the same as the one you first saw. In physics, this effect is known as the Heisenberg principle – you can't interact with a system without changing it. There's a Heisenberg principle in business, too: it's the way you change a game by joining it. That's where we begin this chapter.

To show how this works in practice, we'll examine three different stories of becoming a player. Two of the newcomers fared poorly, one made a lot of money. That was the one who understood how its entry would change the game. We'll draw out the general lessons from these stories. With these lessons in mind, you'll know how to ensure that the decision to become a player is a profitable one.

1 Becoming a Player

Generally, if you want to play, you have to pay. The cost of becoming a player can be cheap when, for example, it means quoting a price over the phone. It's more expensive if you have to pitch an advertising or marketing campaign. It's more expensive still if you have to hire consultants, lawyers, and bankers in order to make a takeover bid. And it can be *really* expensive when you have to build a specialized plant. That was the case for the Holland Sweetener Company, which, back in the mid-1980s, built a $50-million plant to make aspartame.[1]

Bittersweet Success Aspartame is a low-calorie, high-intensity sweetener much better known by Monsanto's brand name for it, NutraSweet. It was the key to the explosive success of diet Coke and diet Pepsi in the 1980s. For people looking to cut calories, it's a godsend:

> **A tenet of western culture is that there is**
> **no pleasure without a price.**
> **What we are saying is that there is a free lunch.**
> — Bob Shapiro, CEO, NutraSweet[2]

The lunch may have been free of calories, but that was the only sense in which it was free. NutraSweet made over half a billion dollars in 1985. The business had 70 percent gross margins. Such profits usually attract entry, but NutraSweet was protected by a patent. What would happen when the patent expired?

High-intensity sweeteners have a long and checkered history. In Roman times, grape juice was boiled down in lead pans to produce sapa, a sweet compound used for everything from a food additive to an oral contraceptive. Unfortunately, the lead in sapa made it hazardous, even lethal. Cyclamate was discovered in the 1960s but was banned in 1970 by the Food and Drug Administration (FDA) after studies suggested a link to cancer. In the United States, the only alternative to aspartame was saccharin, a petroleum derivative discovered back in 1879. In 1977 the FDA tried to ban saccharin, too, as carcinogenic. But the public protested, Congress intervened, and saccharin is still on the market. Apart from the safety issue, some people find saccharin to have a slightly bitter, metallic aftertaste.

Aspartame was discovered by accident. In 1965 James Schlatter, a research scientist at G. D. Searle & Co., was trying to develop an anti-ulcer drug. While experimenting with L-aspartic acid and L-phenylalanine, Schlatter noticed a sweet taste when he happened to lick his finger. He later coined the term 'aspartame' for the combination of amino acids. Aspartame has the same caloric content as sugar of equal weight but is 180 times as sweet.

In 1970 Searle secured a patent on aspartame and sought FDA approval for the use of aspartame as a food additive. When the FDA granted permission for dry use of aspartame in July 1981, Searle quickly launched its first aspartame product, the tabletop sweetener Equal. Use in soft drinks wasn't approved until July 1983. Following these long delays, Searle managed to get the use patent extended – to 1987 in Europe and 1992 in the United States.

In 1985 Monsanto acquired Searle and, with it, ownership of the aspartame monopoly. For Monsanto, this was coming full circle. Today Monsanto is a major producer of agricultural and chemical products, but its original mission, back in 1901, was to challenge a German monopoly on the saccharin market.

What goes around comes around. In 1986 the Holland Sweetener Company began building an aspartame plant in Geleen, the Netherlands, to challenge Monsanto's hold on the aspartame market. Holland Sweetener was a joint venture between two chemical companies, the Japanese Tosoh Corporation and the local DSM (Dutch State Mines). It was created with the express purpose of challenging Monsanto's monopoly of the aspartame market. The process of making aspartame is quite complicated, so Holland Sweetener didn't expect a flood of other entrants when Monsanto's patent expired.

With the expiration of NutraSweet's European patent in 1987, Holland attacked the European market. Monsanto fought back with aggressive pricing. Before Holland's entry, aspartame prices had been $70 per pound. After Holland's entry, they fell to $22–$30 per pound. Holland was losing money. To survive, it appealed to the European courts, which imposed antidumping duties on Monsanto.

Having survived the war in Europe, Holland Sweetener was now prepared to go after the big prize. As Ken Dooley, Holland Sweetener vice-president of marketing and sales, put it: '[We're] looking forward to moving the war into the US.' He had been getting a warm reception from US soft drink companies. 'Every manufacturer,' he pointed out, 'likes to have at least two sources of supply.'[3]

But the war was over before it began. Just prior to the US patent expiration, both Coke and Pepsi signed new long-term contracts with Monsanto. When, at last, there was real potential for compe-

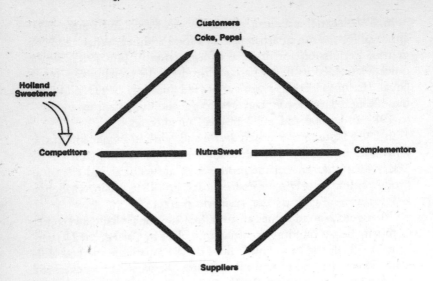

tition between suppliers, it appeared that Coke and Pepsi didn't seize the opportunity. Or did they?

Neither Coke nor Pepsi ever had any real desire to switch over to generic aspartame. Remembering the unfortunate result of the New Coke reformulation of 1985, neither company wanted to be the first to take the NutraSweet logo off the can and create a perception that it was fooling around with the recipe. If only one switched over, the other would most certainly make a selling point of its exclusive use of NutraSweet.[4] After all, NutraSweet had established a reputation for safety and good taste – and not by accident. Searle, and then Monsanto, made huge investments in creating the brand identity. They gave discounts of up to 40 percent to manufacturers who agreed to use 100 percent NutraSweet and to display the distinctive red and white NutraSweet 'swirl' logo on their products. This was backed up with extensive consumer advertising. By 1986, 98 percent of American diet-soda drinkers who used 'artificial' sweeteners recognized the swirl logo.[5] Even though generic aspartame would taste the same, most consumers would be unfamiliar with the unbranded product and would likely view it as inferior.

Another reason not to switch was that Monsanto had a significant cost advantage. It had spent the previous decade marching down the learning curve – cutting its manufacturing costs by 70 percent – while Holland Sweetener was still near the top. With its investment in branding, advertising, and cost reductions, Monsanto appears to have followed the biblical lesson: use the seven years of plenty to prepare for the seven years of famine.

So what Coke and Pepsi really wanted was to get the same old NutraSweet at a much better price. This they accomplished. The new contracts led to combined savings for Coke and Pepsi of $200 million annually.

That was the predictable outcome. Compare Monsanto's bargaining position before and after Holland Sweetener entered the game. Beforehand, Monsanto was in an extremely strong position. Coke and Pepsi had no good alternative to NutraSweet. With cyclamates banned and saccharin under a cloud, NutraSweet made a safe, good-tasting low-calorie drink possible. Its added value was huge. When Holland Sweetener came along, NutraSweet's added value was substantially reduced. NutraSweet's added value was now based on the comparison with generic aspartame rather than with saccharin. What was left was NutraSweet's brand equity and manufacturing cost advantage.

Where did this leave Holland Sweetener? Its entry into the game substantially reduced NutraSweet's added value, and that was worth a lot to Coke and Pepsi. Thus it would have been quite reasonable for Holland, before entering the game, to demand compensation – perhaps a fixed payment or a guaranteed contract – for coming in. But it was much harder for Holland to make money actually playing the game. It had an unbranded product and higher production costs than NutraSweet. Holland didn't have any added value. The pie would have been no smaller without Holland – it just would have been divided up differently. Dooley was right when he said that all manufacturers want a second source. The problem is that they don't necessarily want to do much business with it.

Coke and Pepsi did well to change the game by encouraging the entry of a new player, thereby reducing their dependence on NutraSweet. Monsanto did well to create a brand identity and cost advantage, thereby minimizing the negative effects of entry by a generic brand.

As for Holland Sweetener, perhaps it was too quick to become a player. Holland could save Coke and Pepsi $200 million a year; what could Coke and Pepsi do in return? Holland Sweetener was in a very weak position when it came to selling aspartame – it had no added value. But when it came to selling *competition*, Holland was in a very strong position. In a sense, it had a monopoly: it was the only one Coke and Pepsi could use to improve their bargaining position with NutraSweet. But Holland gave that service away. Perhaps Coke and Pepsi would have paid for this valuable service, but only if Holland had demanded payment up front. In the upcoming section 'Pay Me to Play,' we'll suggest a variety of ways of getting paid to enter a game.

Keep track of the Holland Sweetener story as you evaluate Norfolk Southern's strategy in its negotiations with Gainesville Regional Utility.

Another Coal Porter In Gainesville, Florida, electricity and drinking water come from the city-owned Gainesville Regional Utility.[6] This utility has a problem. It has always been dependent on one railroad, CSX, for delivery of all its coal. As you might expect, CSX has done well selling coal to Gainesville: in 1990 the price was $20.13 a ton. So it was quite a coup that July when Gainesville negotiated a deal with Norfolk Southern railroad to deliver coal at $13.68 a ton.

The only problem was that Norfolk Southern couldn't quite deliver because its closest line came up short – twenty-one miles from Gainesville. Close wasn't good enough.

Gainesville asked CSX for permission to let Norfolk Southern use its tracks. CSX declined. Why should it give away its monopoly? That was just Gainesville's opening move. With the option for the cheaper coal, the city decided it was worth building a twenty-one-mile connecting line, even at a projected cost of $28 million.

At this point, the situation looked promising for Norfolk Southern. It had put Gainesville on the line, so to speak, for the costs of building the spur, and it had a contract to deliver coal at $13.68 a ton if the line was in fact built. No downside, only upside.

Norfolk Southern did have to expend some political capital. Because the proposed line would pass through environmentally sensitive wetlands, Gainesville and Norfolk Southern had to go before

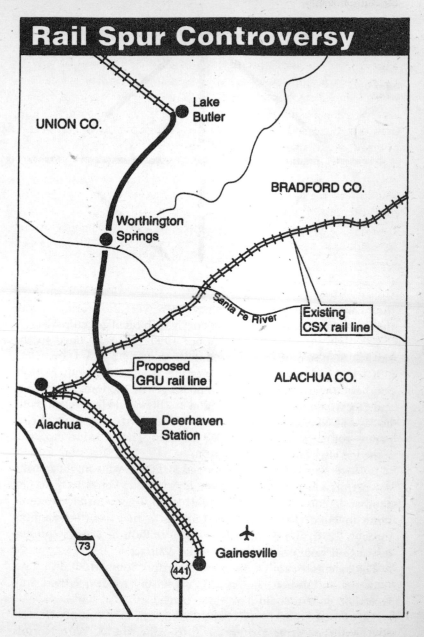

Rail Spur Controversy

UNION CO.

Lake Butler

BRADFORD CO.

Worthington Springs

Santa Fe River

Existing CSX rail line

Proposed GRU rail line

ALACHUA CO.

Alachua

Deerhaven Station

73

441

Gainesville

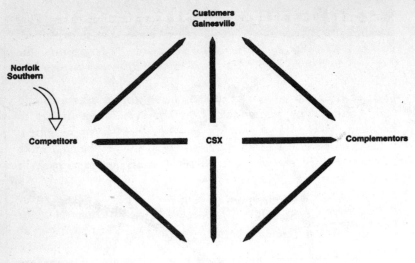

the Environmental Protection Agency. There were also hearings before the Interstate Commerce Commission, which regulates railroads.

CSX didn't stand still. In October 1991 CSX responded to the Norfolk Southern offer by lowering its price by $2.25 a ton. Then CSX let it be known that if Gainesville went ahead with Norfolk Southern, the existing line would no longer be economically viable, and CSX would be forced to abandon it. This would leave Gainesville hostage to Norfolk Southern. Donna Rohrer, vice-president of corporate communications at CSX, warned: 'They'll wind up with someone else, but not competition.'[7]

In November 1992, when it looked as if Gainesville might actually get permission to build the spur line, CSX lowered its price by another $2.50 a ton. At that point, it no longer made economic sense to build the connection. CSX got a new contract running through 2020, and Gainesville got $34 million in savings (present discounted value) over the life of the contract.

The outcome wasn't a disaster for Norfolk Southern. It didn't get Gainesville's business, but it didn't lose any money, either. Still, Norfolk Southern could have done better.

The mistake was to give away competition. Without Norfolk Southern, Gainesville had no threat to make to CSX. With Norfolk

Southern, the city could and did make a very credible threat. That was the game. As Anthony Hatch, a PaineWebber analyst, explained: 'It's an expensive venture to build a railroad ... Sometimes, the threat to do it is enough.'[8] Here, making the threat was worth $34 million.

To ensure itself a piece of the action, Norfolk Southern should have required a fee, perhaps based on Gainesville's cost savings, in the event that the spur line didn't get built. Then Norfolk Southern would have been in a no-lose position: if the line gets built, Norfolk Southern gets the business; if it doesn't, Norfolk Southern shares in the cost savings it made possible.

Even though it wasn't out-of-pocket, Norfolk Southern did incur some hidden costs. It used up some political capital in lobbying to get the spur approved. Its actions also imposed a $34-million cost on CSX – quite dangerous given CSX's ability to retaliate. In the whole of the United States, only twenty or so power stations are served by more than one railroad. Just as CSX had the only tracks into Gainesville, there are many other places where Norfolk Southern is the monopoly supplier and CSX can play the spoiler.

Norfolk Southern should have known better. It had already been on the receiving end of this game twice before. Up until 1991, the Southern Company, an Atlanta-based utility, had been served exclusively by Norfolk Southern.[9] That was until it built a seven-mile line from a plant near Birmingham, Alabama, to a track owned by CSX. This was the first time a utility had ever built a rail access line for the specific purpose of creating competition. Similarly, in Evansville, Indiana, PSI Energy threatened to build a ten-mile rail extension from its plant to a CSX main line. Once again, Norfolk Southern was the monopoly incumbent. It was forced to give CSX access to its tracks, thereby creating a competitor to itself, at which point PSI gauged it unnecessary to build the rail extension.

Perhaps what Norfolk Southern did in Gainesville was a case of tit-for-tat retaliation: after CSX spoiled Norfolk Southern's monopoly with the Southern Company and PSI Energy, Norfolk Southern spoiled CSX's game in Gainesville. But that's unlikely to be the final chapter. There's a real risk that CSX will retaliate. So Gainesville wasn't no downside, only upside, for Norfolk Southern. Gainesville was really more like 'Loseville.'

As the Holland Sweetener and Norfolk Southern stories illustrate, sometimes the most valuable service you can offer is creating competition, so don't just give it away. That's especially true when providing competition is costly: Holland had to build a multimillion-dollar plant, and Norfolk Southern's bid risked provoking a tit-for-tat response. Competition is too valuable and too costly to give away. You need to get paid to play. This idea is well understood by players in the takeover game.

Answering the Call The cellular phone business was undergoing rapid consolidation when, in June 1989, thirty-nine-year-old Craig McCaw made a bid for LIN Broadcasting Corporation. Five years earlier, the Federal Communications Commission had divided the country up into 306 separate markets and allocated two cellular licenses to each one. One of those licenses had been earmarked for the local phone company, while the other had been awarded via lottery. McCaw had been going around and buying up licenses from the lottery winners. To date, McCaw's licenses covered 50 million potential customers – or POPs in the industry jargon.

Already the industry leader, McCaw wanted to go national. The acquisition of LIN's 18 million POPs was McCaw's best, and possibly only, chance to gain major city franchises and thereby create a national cellular footprint. That meant he had to get LIN, which had cellular licenses for New York, Los Angeles, Philadelphia, Houston, and Dallas. McCaw already owned 9.4 percent of LIN. Now he wanted the rest.

McCaw had a vision and had already demonstrated his willingness to roll the dice. He had taken on a huge amount of debt in order to buy all the licenses. Yet, as of mid-1989, only 250,000 out of his 50 million POPs were actually paying customers – a penetration of one-half of 1 percent. McCaw saw the glass as nearly full as opposed to almost empty: the *potential* profits were enormous.

McCaw bid $120 a share in cash for LIN, a total of $5.85 billion. His bid resulted in an immediate jump in LIN's share price from $103.50 to $129.50. Clearly, the market expected more action. But there was a catch to McCaw's bid: the offer was conditional on LIN's removing its 'poison pill' antitakeover defense. LIN refused to do so. Donald Pels, LIN's CEO, had a long-standing aversion to

Craig McCaw, and the feeling was no doubt mutual.[10] Should McCaw succeed, the chances that Pels would keep his job were slim to nonexistent. Faced with a hostile reaction and the poison pill, McCaw lowered his offer to $110.

LIN sought other suitors. Several of the Baby Bells were rumored to be interested. BellSouth, with its vast financial resources and a strategy of buying cellular, seemed the most likely to answer the call. LIN's Los Angeles and Houston franchises would complement BellSouth's existing holdings in those cities. The combined holdings of the two companies would come to 46 million POPs, only 4 million short of McCaw's 50 million – not quite a national footprint, but a very serious threat to McCaw's lead position.

For BellSouth, the chance of winning a bidding war against McCaw was low. McCaw was projecting that 15 percent of Americans would use cellular phones by the year 2000, making each POP worth about $420. The consensus forecast among the Baby Bells was that penetration would be only 10 percent. That led to valuations of only $280 per POP. In a contest between BellSouth and McCaw, the smart money was on McCaw.

McCaw had an extra motive to bid high. A rise in the market valuation of POPs would make the 50 million POPs he already owned all the more valuable. For BellSouth, the cost of playing the game was high: attorney and investment banker fees could top $20 million, and top company executives would find themselves distracted from their job of running ongoing operations. Moreover, if it lost, as was the likely outcome, BellSouth would lose credibility as a determined bidder in the future. Egos would get bruised.

Nevertheless, BellSouth was willing to acquire LIN for the right price. The problem, of course, was that by entering the fray, it would trigger a bidding war, and then LIN wouldn't sell for a reasonable price. BellSouth knew only one bidder could win, and it wanted something for its trouble in the likely event that the winner was McCaw. Thus, as a condition for bidding, BellSouth got LIN's promise of a $54-million consolation prize and an additional $15 million toward expenses should it be outbid. Having been paid to play, BellSouth made an offer that market analysts valued at $105–$112 per share.

As expected, McCaw did not give up. He outbid BellSouth with a new offer valued at $112–$118 per share.[11] LIN came back to BellSouth and asked it to up its bid. In return, BellSouth again asked to be paid. LIN raised BellSouth's expense cap to $25 million, and, in return, BellSouth entered a new bid valued at $115–$125 per share.

McCaw raised his offer to an amount valued at $124–$138 per share and then added a few dollars more to close the deal. At the same time, he paid BellSouth $22.5 million to exit the game.[12] At the final price, LIN was valued at $6.3–$6.7 billion. At this point in the bidding, Pels recognized that his stock options were worth over $100 million, and the now friendly deal with McCaw was done.

So how did the various players fare? The day before BellSouth announced its first bid, LIN was trading at $105.50. When BellSouth exited, the stock price was $122.25. The net rise of $16.75 a share comes out to almost $1 billion. So LIN got itself an extra billion, which made the $54 million plus expenses it paid BellSouth look like a bargain. McCaw got his national network and subsequently sold out to AT&T, making him a billionaire. And BellSouth, by being paid first to play and then to go away, turned a weak hand into a $76.5-million fee plus expenses. BellSouth clearly understood that even if you can't make money in the game the old-fashioned way, you can get paid to change it.

With the benefit of hindsight, a few questions come to mind. Why didn't BellSouth get the billion instead of $54 million? Perhaps if BellSouth had been too greedy, LIN might have turned to another Baby Bell. LIN might even have agreed to pay it a larger consolation prize, but that would have been very risky. LIN would have become prohibitively expensive for anyone else to acquire, and then the courts might have invalidated the fee as a 'lockup,' in which case BellSouth would have gotten nothing.[13]

Why did McCaw pay BellSouth to exit? McCaw could not be sure BellSouth would exit on its own or, if it did exit, whether it would be sooner or later. Sooner would be better. McCaw didn't want any unexpected new bidders. As a senior executive at McCaw put it: 'We nudged BellSouth to just walk away with $26 million.'[14]

McCaw's payment to BellSouth to stop bidding was exotic but perfectly legal. American antitrust law generally prohibits one bidder from paying another to exit an auction. However, the takeover of a publicly traded corporation falls under securities law, and courts have held that securities law preempts antitrust law.[15] Securities law does not prohibit 'exit fees'; it only requires full disclosure. The thinking is that the prospect of getting an exit fee encourages weaker bidders to come into an auction in the first place and thereby increase share prices. Once several bidders have entered, shareholders of the company in play would obviously like to disallow exit fees. But that's like trying to have it both ways.

Why didn't LIN insist on a no-exit-fee clause in its contract with BellSouth? Perhaps LIN didn't think of it. Still, it's curious. If LIN thought to encourage BellSouth to enter with a payment, it should have considered that someone else might pay BellSouth to go away. It could have prevented that from happening.

We've now seen three different stories of becoming a player. Holland Sweetener and Norfolk Southern both fared poorly. Other players benefited at their expense. BellSouth did much better. The reason it made money is that it recognized who stood to gain from its entry. Having recognized this, it could negotiate for a share of that benefit.

A prospective player should always ask Cicero's question: *Cui bono?* Who stands to gain? Holland Sweetener and Norfolk Southern didn't ask Cicero's question. BellSouth did.

Let's apply the lesson of these stories to a familiar, everyday business situation.

Pay Me to Play The phone rings. A prospective customer explains that he is not satisfied with his current supplier and would like you to give him a bid. This is a large account that's been with one of your competitors for some time. Here's your chance.

What do you do? First, you tell the caller on the other line that you'll have to get back later. Then you ask the customer for some specifics that will help you price the bid. When you get off the phone, you assemble a team to work out a bid. In spite of the friendly call, you know it's a long shot, and that prompts you to work just a little harder and be just a little more aggressive on price.

In the back of your mind is the suspicion that the customer is using you to get a better price from his current supplier. But that's the way the game is played, isn't it? If you don't bid, there's no chance of getting the business. Moreover, you risk alienating the customer and losing any chance of future business. And how do you explain to your boss that you passed up an opportunity to get this business, especially if it turns out that the customer does end up switching suppliers?

So you make an aggressive bid. It seems that not making a bid has no upside and a real potential downside, while making a bid has no downside and a chance of an upside. The customer thanks you and promises to get back to you. But he doesn't.

What might you have done differently? You could have bid lower, but there's no guarantee that would have worked any better. That's not a real solution.

The problem with the strategy was more basic. Your aggressive bid no doubt helped the customer get a lower price from someone else. That was the likely outcome all along. The customer ends up better off, and you have nothing to show for it.

There seems to be a natural impulse to offer competition for free. After all, that's what business people are supposed to do, isn't it? You want a bid? I'll give you a bid. Contractors, architects, exterminators – they all give bids. So, in effect, did Holland Sweetener and Norfolk Southern. But BellSouth didn't, at least not for free.

The right question to ask is: how important is it to the customer that you bid? If your bidding is important, then you should get compensated for playing the game. If it's not so important, then you're unlikely to get the business and even less likely to make money. You might want to reconsider bidding at all.

That's nice in theory, but does it work in practice? Everyone would like to be BellSouth and get paid $76.5 million for playing a losing hand. But that's pretty rare. Most customers would simply laugh at you – more likely swear at you – if you tried to get them to pay you in cash for making a bid.

Competition is valuable.
Don't give it away.
Get paid to play.

It's abrasive to ask to be paid in cash; it's often not that smart, either. Fortunately, there are many other ways of getting paid to play. You can ask for contributions toward bid-preparation expenses. You can ask for help with up-front capital costs, such as the costs involved in building a plant. You can ask for a guaranteed sales contract. Also valuable is a last-look provision: you get the business so long as you match the best price in the market.

In return for bidding, you can ask for better access to information about the business. That gives you a much better chance of winning the account. It turns you from being an outsider to being an insider. It's the first step toward forming a relationship with the customer.

Ask to deal with a different person. Make the bidding an opportunity to meet senior management. Ask to meet someone who will appreciate what you bring to the table and not just focus on getting the lowest price. Or in return for bidding on one piece of the business, get access to the customer's other pieces.

Finally, you might try turning the tables. Instead of quoting the customer a price, ask the customer to quote you a price at which he would give you his business. The customer gives you a signed contract complete with price, and you decide whether to sign. If you do, the customer has to switch to you. This way, you can ensure that the customer isn't playing games. Car dealers know this technique quite well. Instead of quoting you a price, they often ask you how much you'd pay. After hearing your price, they say: 'I'd like to sell you the car at that price, but I need the manager's approval.' But first, they need you to sign a contract at the price you've named. Dealers know, then, that if they agree to your price, they will have a sale and not just another round of negotiation.

SEVEN WAYS TO GET PAID TO PLAY

1. Ask for contributions toward bidding expenses, up-front capital costs, and other costs of entering the game.
2. Ask for a guaranteed sales contract.
3. Ask for a last-look provision.
4. Ask for better access to information.
5. Ask to deal with someone who will appreciate what you bring to the table.
6. Ask to bid on other pieces of business, in addition to the current contract.
7. Ask the customer to quote a price at which he would give you his business.

Cash is okay, too.

Of course, even with this arsenal of strategies, you might still fail to get paid, in which case you may decide not to play. That outcome isn't as bad as you might think. The cost-benefit calculation that suggested you might want to become a player in the first place was a little off. It's simply not true that making a bid has no downside.

Eight hidden costs of bidding

There are hidden costs associated with making a competitive bid.

1. There are better uses of your time. Making a bid typically takes a lot more time and effort than simply reading a number off a price sheet. That time and effort often takes priority over serving your current customers. Keeping your current customers happy is smarter than chasing after other people's customers.

2. When you win the business, you lose money. If you win the business, you should be a little suspicious. Do you really want to win a customer just because of your low price? No. A customer you win on price alone is telling you he has no loyalty. If you think that getting the customer gives you an opportunity to make money later by raising price, think again: by coming to you, this customer has just revealed himself as someone who will switch suppliers in order to get a lower price. Thus, you'd better make sure that you can make money at the price you offer to attract the customer.

Ask yourself: why is the incumbent letting the customer go? Perhaps the customer doesn't pay his bills. Perhaps he's particularly demanding. If this were a good customer to have, your competitor would have kept him. The fact that you were able to steal the customer away should give you pause for thought.[16]

Sometimes there is a legitimate explanation for attracting the customer. Perhaps your rival really messed up and that's why the customer is leaving. But in that case you don't need to use a low price to attract the customer.

In sum, it's hard to attract someone else's customer away by using price *and* then make money at that price. Only if you have a lower cost structure can you afford to undercut the incumbent and still make a profit.

3. The incumbent can retaliate. Don't think that winning this customer is going to be the end of the game. If this is a good customer, then your win is someone else's loss. (If it's a bad customer, then you've already made a mistake.) The incumbent supplier is quite likely to respond. He can go after one of your customers. He may not get your customer, but he can surely force you to lower the price. If he succeeds in snagging your customer, then you and your rival have turned two high-margin accounts into two low-margin accounts. And you've traded accounts with a relationship for accounts in which a relationship needs to be established. Even if your rival fails to snag a customer of yours the first time, he might try again and continue doing so until he succeeds. The end result: lose-lose.

4. Your existing customers will want a better deal. The price you offered to get the new account is unlikely to stay secret. If your current customers find out how low you are willing to go to get a new account, they will likely demand that you offer them at least as good a price. They may even have contractual provisions guaranteeing them the best price that you offer anyone else. The result of going after the new account, whether or not you get it, is that your existing customers may now have a reasonable case to get you to give them a lower price. That can be a very high cost.

5. New customers will use the low price as a benchmark. The bad precedent goes beyond givebacks to your existing accounts. Think ahead to the next time a new account – someone who is new to the business and has no incumbent supplier – comes knocking at your door. The low price that you offered this time becomes a benchmark in the bidding for the new customer.

6. Competitors will also use the low price as a benchmark. Even if you were willing to risk charging a higher price again in the future, your rivals might expect you to come in with a low price, and these expectations become a self-fulfilling prophesy.

In sum, existing customers, future customers, and your competitors will all use the low price as a benchmark in the future.

7. It doesn't help to give your customers' competitors a better cost position. Your future and that of your customer are naturally linked. If your future is tied to Coke, you don't want to help Pepsi get a lower price. Unless you have very good reason to believe that you can get Pepsi's business and keep Coke's, bidding for Pepsi's business is costly. You help your competitor's customer and thereby hurt your own.

8. Don't destroy your competitors' glass houses. Lowering your competitor's profits isn't necessarily smart. You're entitled to get concerned if a rival is building a war chest. But that doesn't mean you should aim to deflate his profits – just as the FedEx messenger shouldn't let the air out of the UPS truck's tires. That would be absolutely, positively a bad idea.

The view that you win if competitors lose is simplistic and potentially dangerous. If you lower your rival's profits, he then has less to lose and every reason to become more aggressive. He can go after your existing accounts with abandon. In contrast, the more money your rival is making, the more he has to lose from getting into a price war. Until your rivals live in glass houses, expect them to throw stones. Thus, it's in your interest to help them build a glass house – not a mansion, a house.

EIGHT HIDDEN COSTS OF BIDDING

1. You're unlikely to succeed – there are better uses of your time.
2. When you win the business, the price is often so low you lose money.
3. The incumbent can retaliate – you end up trading high-margin for low-margin customers.
4. Win or lose, you help establish a lower price – your existing customers will then want a better deal.
5. You set a bad precedent – new customers will use the low price as a benchmark.
6. Competitors will also use the low price you helped create as a benchmark.
7. It doesn't help to give your customers' competitors a better cost position.
8. Don't destroy your competitors' glass houses – if they're a little vulnerable, they'll be less likely to go after your accounts.

It's very tempting to just go ahead and bid when you're asked to. Now you have some reasons to look before you leap, or not leap at all. And if your boss asks why you weren't willing to play the game for free, you can explain about the Eight Hidden Costs of Bidding.

The flip side of entering a game is having someone else enter your game. What do you do then?

Calling All Players The phone rings. Once again, it's a customer who calls to say that he's unhappy with his current supplier. Only this time, that means you: you're his current supplier. You inquire why he's upset, and he tells you that someone else has come in and offered to supply the same service for 50 percent less. He asks: 'What are you going to do?'

Good question. You take a deep breath. Next you have to ask yourself whether the customer is really telling the truth.

It's not unheard-of for people to make up bids in order to get incumbent suppliers to come up with better prices. It's a cheap trick. It's neither ethical nor, in the long run, effective. It puts the supplier in a lose-lose position. Matching a fictitious bid is giving away money, but calling the bluff is also a loser. Once the customer has been exposed as a liar, it's virtually impossible for supplier and customer to go on doing business together – the cat has been let out of the bag. (More on this in the Tactics chapter.) What about calling up competitors to check whether the bid is real? In the United States, at least, that's illegal.

Okay, you have to assume that the customer is telling the truth. Now what? Go ahead and match the lower bid? That may not be necessary. First, you should remind the customer of your track record, assuming, of course, that it's a good one. Emphasize that in switching suppliers, he'd be giving up a proven relationship for a leap into the dark. Lower price or not, he could well end up regretting the move. If the customer values the relationship, you should be able to keep the business without meeting the new price. You'll probably need to make some price concession, but you won't need to go the whole way.

If this doesn't work, that doesn't mean you should now go ahead and match price. It might be better to let the customer go. Matching price can be very expensive: your other customers may get wind of the offer and demand similar discounts. You may even be contractually obligated to give them the same lower price.

As for letting the customer go, that may be a hard bullet to bite, but you may have to do it only once. This may be a case where your competitor has a little unused capacity – just one bullet in his gun – and will keep aiming it at different targets until he connects. If you fight and succeed in keeping the customer, the battle will just shift to another of your customers, and so on. You could fight all the way and deny your rival any of your customers, but there would go your profits. A Pyrrhic victory.

Perhaps you should let your rival shoot his bullet and be done. But if conceding this account will lead your competitor to become more aggressive in going after your other accounts, you should work to keep the customer, even at some cost. To decide what to do, you have to ask what the competitor will do if he doesn't get the customer, and what he'll do if he does.

So far we've looked at entering a game, and also at how to respond if someone – an unwelcome guest – enters your game. Now we turn to the strategy of bringing other players into the game.

2 Bringing in Other Players

Players in a game often want to expand the cast of players. We've already seen some examples of this. LIN brought in BellSouth as a second bidder, and Gainesville brought in Norfolk Southern as a second supplier. Coke and Pepsi would, no doubt, have been prepared to pay Holland Sweetener handsomely to become a second supplier.

We'll start this section by looking at the strategy of bringing in customers. Then we'll turn to bringing in suppliers. And we'll go tour the rest of the Value Net. You certainly want to bring in complementors, and in some cases, even bringing in a competitor can be beneficial.[17]

Bringing in customers

It's a good idea to bring more customers into the game. One benefit is obvious: it's a bigger pie. More customers lead to more sales, which, in turn, lead to more profits. There's another benefit. With more customers, no one customer is as essential. Bringing in new customers lowers the added values of all the existing ones. That puts the seller in a stronger bargaining position with respect to its customers. So, for the seller, it's a double win: the pie grows and he gets a bigger share.

Think back to the Card Game. Barry did well to lose three black cards. His bigger slice more than made up for the smaller pie. He would have done even better if, instead, he had found three extra red cards and handed them out to the students. That is, instead of playing a game with twenty-three black cards and twenty-six red cards, Barry plays a game with twenty-six black cards and twenty-nine red cards. Once again, no individual student has any added value. But, in this case, the pie grows rather than shrinks. Barry ends up capturing the lion's share of the bigger pie.

That's all fine if you have one side of the market all to yourself, as Barry did. But what if you face competition? You bring more buyers into the game, but they don't necessarily belong to you. It's expensive to develop the market and drum up demand, so why do it if your competitor might be the beneficiary?

Loghandlers at Loggerheads Harnischfeger Industries is a Milwaukee-based engineering company that makes portal cranes.[18] And just what are portal cranes? Forest products companies, such as Georgia-Pacific, International Paper, and Weyerhaeuser, traditionally moved logs around in their woodyards with mobile log stackers – diesel electric vehicles somewhat akin to giant forklifts. In the mid-1970s large specialized portal cranes, which have giant grapples (like claws) and move logs around by picking them up from above, began to replace the stackers.

Portal cranes allow a woodyard to move logs around more efficiently. In theory, if Harnischfeger could have captured all the cost savings, it could have made about $5 million a crane. The catch was competition – which came in the form of entry in 1987 by Kranco, a small crane maker that was a leveraged buyout headed up by several former Harnischfeger executives. Not surprisingly, Kranco's product and cost position closely matched Harnischfeger's.

With buyers coming into the market in only a trickle – usually one at a time – each was able to play Harnischfeger and Kranco off against the other. By soliciting bids from both crane makers, a buyer was able to capture the lion's share of the $5 million of value. That was the problem facing Harnischfeger.

What options did Harnischfeger have for changing the game? One possibility was the classic win-lose approach: kill the competition. Kranco was a cash-hungry leveraged buyout while Harnischfeger had deep pockets. An extended price war might have starved Kranco to death. In the meantime, though, Harnischfeger would have starved itself, too. There was a better way to change the game: find more buyers.

Instead of competing for a small set of existing buyers, Harnischfeger could have worked to create some new ones. The cost savings from a portal crane come only if a woodyard is configured to process tree-length logs. But the vast majority of woodyards had been designed before tree-length technology came along: they used short logs, and so portal cranes offered no savings. Harnischfeger could have expanded the market dramatically by showing woodyards the benefits of the new tree-length technology.

But what if all these new buyers had purchased from Kranco? Then Kranco would have been the big winner. Wouldn't that have been a problem? No. It would have been win-win. With more customers of its own, Kranco would have been less desperate to go after each and every customer that approached Harnischfeger. And buyers couldn't have played the two sellers off against each other. With each customer, Harnischfeger and Kranco would have been likely to capture more of the $5-million savings.

No more price war. Kranco didn't have to lose for Harnischfeger to win.

In fact, the win-win would have been a big win for Harnischfeger. Kranco had only limited crane-building capacity. It couldn't have absorbed that many new buyers. This is one of those cases where the competitor had only a few bullets in its gun. And once those few bullets were spent, Kranco could do no more damage.

What actually happened? Harnischfeger opted to continue the price war – a win-lose approach that ended up lose-lose. Kranco declared Chapter 11. But it didn't disappear. It was acquired by Kone, a leading Finnish engineering company, and today Harnischfeger faces a more formidable competitor.

Elements of the Harnischfeger story can be found in many other businesses. Take airplanes, for example. Orders for new planes are large and infrequent, and so the airframe manufacturers, Boeing and Airbus, view each one as a must-win. Commercial airlines are able to play Boeing and Airbus off against each other. Anything that either manufacturer could do to bring a few more buyers into the game would make a big difference. It's even okay for Boeing if those new buyers go to Airbus. The reason is that there's limited manufacturing capacity. If Airbus wins several consecutive orders, it starts getting a large backlog. Now Boeing will be able to promise faster delivery and thus be better positioned to win the next few orders. If there are only a few buyers, not enough to create a backlog, then Boeing can't afford to let Airbus win one. Every lost order puts more pressure on Boeing's overhead. Competition heats up until neither Boeing nor Airbus makes money. Just a small shift in the number of customers, one way or the other, can make a big difference to the balance of power in the market. It's the Card Game again. Just a small change in the number of black – or red – cards is all that it takes to shift the balance of power in the game.

Bringing more customers into the game is a good idea. That's true when you have no competition and can be even more important when you do. As for how to bring in customers, we've already explored some options. One way is to educate the market, as Harnischfeger might have done. Another is to pay them to play, an idea we explored in the first part of this chapter.

Sometimes, paying customers to play, especially early adopters, is essential. You need to get the ball rolling. The classic example is selling a network service. The more people that use America Online, the more everyone values it because that's where they'll find more people on-line. Likewise, the more people who have ProShare, the Intel videoconferencing system we discussed in the Co-opetition chapter, the more valuable it becomes to everyone. There are more people to call and more people who can call you.

America Online understood that it had to lose money early on – effectively paying people to play – in order to build up a customer base. Likewise, Intel is subsidizing ProShare. The good news is that they don't have to pay everyone to play; once there's a critical mass of customers, others will follow of their own accord.

Getting a party going isn't very different from starting a network. No one wants to go to an empty nightclub, so nightclubs often start out with free admission, even free drinks, early in the evening to get things started. Nightclubs pay some people to play to ensure that the rest will pay them to play.

Newspapers and magazines also subsidize some of their customers. That's because they have two sets of customers: readers and advertisers. The greater the number of readers, the more advertisers are willing to pay. To boost circulation, publishers usually sell their papers below cost. The increased advertising revenue more than offsets the subsidy to readers. Many publishers would be willing to give their product away, or even pay people to take a copy – but only if they could be sure people would read it. The problem is that if the price gets too low, subscribing is no longer evidence of any real interest or commitment, and advertisers worry that the publication will be thrown away without being read.

Yet another way to bring in more customers is to identify and stimulate complementary products. Developing complements naturally attracts more customers into the game. We introduced this idea in the Co-opetition chapter, and we'll talk more about it below, in the 'Bringing in Complementors' section.

Finally, consider becoming your own customer.

After World War I, US airplane manufacturers Boeing and Douglas were struggling. There was no more demand for military planes, and civilian aviation had yet to take off. The best opportunity came when the US Post Office put out a bid to deliver airmail. Would the winner buy Boeing's planes or would it buy those of Douglas? Boeing didn't take any chances. It bid for the Post Office contract itself, and won. Boeing built the planes and then created what later became United Airlines to fly them and deliver the mail. To ensure a market for its planes, Boeing essentially created its own captive customer.

Likewise, it's not a coincidence that the automakers own car rental agencies. Ford owns Hertz and part of Budget. Chrysler owns Dollar and Thrifty and has an investment in Avis that requires Avis to buy 20 percent of its fleet from Chrysler. GM also owns part of Avis; in return for GM having taken a 25 percent equity stake, Avis agreed to buy 60 percent of its fleet from GM. For seven years, GM also owned National. Although GM sold the business in April 1995, National's new owners have a long-term agreement to continue buying GM cars. Finally, Mitsubishi owns part of Value Rent-a-Car.

Developing the car rental business helps the automakers sell more cars. It also gives each automaker control over which cars the rental agencies will use. And it's a great way to get people to test-drive the latest models.

Becoming your own customer is a way to develop the market, assure demand, and achieve scale.

BRINGING IN CUSTOMERS

1. Educate the market.
2. Pay them to play.
3. Subsidize some customers, and other full-paying customers will follow.
4. Do it yourself: become your own customer to develop the market, assure demand, and achieve scale.

Bringing in suppliers

Just as bringing in more customers is a good idea, so is bringing in more suppliers. With more suppliers, no one supplier is as essential, and that puts the buyer in a stronger bargaining position. How do you bring suppliers into the game? Just as with customers, one way is to pay them to play. There's another way: form buying coalitions.

Don't Buy Health Care without Us In May 1995 American Express, IBM, ITT, Marriott, Merrill Lynch, Nabisco, Pfizer, and Sears, along with two other large companies that prefer to remain anonymous, formed a buying coalition to purchase the HMO component of their employee health-care benefits. The end result was a pooling of over 600,000 employees and dependents buying $1 billion of health insurance.[19]

American Express had already proved that the coalition concept had merit. In 1994, in conjunction with Merrill Lynch and Macy's, American Express tested the idea in California, Florida, Texas, New York City, and Atlanta.[20] The result was a 7 percent decrease in HMO premiums at coalition sites. In contrast, in places not covered by the coalition, premiums rose by 7 percent.[21]

An obvious benefit of being part of a buying coalition is sheer size. Health-care providers perceive that they can't afford to lose the coalition's business, so they all bid more aggressively.

But that's not all. There is a second – perhaps even more important – source of power. American Express and its coalition partners have attracted over a hundred bidders into the game. As a result, no individual bidder has much added value, and that puts the coalition in an extremely powerful position.

The hundred-plus bidders is many, many more than any of the coalition members could have attracted individually. American Express might attract five, even ten, health-care providers – but not one hundred. The size of the American Express account wouldn't justify the time and effort involved in making a bid – at least, not if there were a hundred bidders. The chances of winning would simply be too small. But with the coalition's billion dollars of business at stake, even a small chance of winning is enough. That's why a hundred bidders found it worthwhile to play.

Even if it could attract a hundred bidders by itself, American Express couldn't do justice to all the bids. It would simply be too expensive. The coalition can share the cost of evaluating the hundred or so different offers. Indeed, there was an exhaustive review of the bidders who made it to the final round. A selection panel interviewed each plan's representative about procedures for handling difficult medical cases. Each candidate had to supply twenty-five medical charts from cases with unfavorable outcomes – such as readmission for the same condition, and even unexpected death.[22]

The American Express buying coalition continues to expand. The ten charter members have been meeting with another ten prospective members to discuss other purchasing initiatives. This combined group – nearly six times as large and representing 3.5 million people – is discussing joint purchasing of mental-health, pharmacy, and other health-care benefits.[23]

Buying coalitions are more prevalent than one might think. Buying clubs such as Price/Costco and Sam's are effectively buying coalitions, as strategy consultant and author Michael Treacy has pointed out.[24] Instead of people going into the local supermarket and deciding which brand of toothpaste or peanut butter to buy, they leave the decision to Price/Costco's professional buyers. These clubs don't offer a large variety, but what they lack in variety, they make up for in price.

Forming buying coalitions is a powerful strategy to bring in more suppliers. We're sure that more companies – independent bookstores, university libraries, and hospitals, to name just three – could profit from taking this approach to purchasing.

American antitrust law does put some constraints on buying coalitions. Coalition members can't be forced to buy with the coalition – the ultimate purchase decisions must be voluntary. Of course, there's usually financial and peer pressure to stay in the coalition. And the coalition must not have 'market power,' which is generally defined as controlling more than 30 percent of the market.[25]

BRINGING IN SUPPLIERS

1. Pay them to play.
2. Form a buying coalition to become a larger buyer.
3. Do it yourself: become your own supplier to assure supply and create competition.

Bringing in complementors

Bringing in customers and suppliers is a good idea. So is the strategy of bringing in complementors. It raises your added value. With more complements, your product becomes more valuable to customers. Even better is if the complements are inexpensive. So you want as many complementors as possible in the game competing with one another. The more the merrier.

How do you bring in more complementors? The idea of a buying coalition is suggestive. If you're bigger than your customers, you can help them do a better job buying from existing complementors. And just as the American Express buying coalition attracted more suppliers into the game, your greater size will attract more complementors into the game and further reduce prices.

We saw an example of this in the Co-opetition chapter with the MacBains and their used-car weekly, La Centrale. The MacBains promise to steer their readers to whoever offers the lowest-priced complements – insurance, financing, and warranties. They have effectively created a buying coalition for their readers and, in this way, can negotiate with more potential providers of complements than their readers could do on their own. The result is that La Centrale's readers save money on complements.

We also mentioned the possibility that auto insurance companies might follow a similar strategy. They could negotiate with dealers to help their customers get better deals on new cars. A few insurers, such as USAA, some credit unions, and AAA (the American Automobile Association), already do this, at least to a limited extent. We think this strategy could be much more widely employed.

Large insurance companies could surely get a better price on new cars than their customers could on their own. While an individual might visit two or three dealers, an insurance company would attract every dealer statewide to bid for its business. An insurance company could go to all the Ford dealers in a state and offer to direct all their customers to any dealer willing to sell at a small markup over cost. Once the insurance company had negotiated this type of deal in one state, it would know what prices were possible and could use that information to help it negotiate in other states.

For car buyers, this would mean no more haggling with dealers. When people wanted to buy a new car, they would just call up the special insurance agent, explain the make and model desired, and the agent would quote the best price and provide the name of the nearest dealers willing to sell at that price. Buying a new car would become a cheaper, quicker, and more pleasant experience. People would buy new cars more often and therefore more auto insurance.

What if insurance companies don't want to get into the car-buying business themselves? They could still implement this strategy by forming an alliance with one of the existing car-buying services, such as Auto-By-Tel, CUC, or Mass Buying Power.

Either way, insurance companies would see a further benefit. Insurance companies aren't always appreciated, to put it politely. People give them real money in return for a promise. Most of the time, they don't see anything tangible in return. And when they do get something, it's because something bad has happened to them, so at best it's a consolation prize. Were they to help their customers buy cars, insurance companies would earn their customers' gratitude.

Bringing more complementors into the game is nice, but sometimes the problem is more fundamental: complements are missing altogether. Video game makers face this problem on a recurring basis. Their next-generation technologies are usually incompatible with the old, so consumers won't buy the new hardware until a critical mass of software exists. Yet software writers won't produce games for the new system until a sufficient hardware base exists. You need the chicken and the egg at the same time.

Cheap Complements The 3DO Company is known for having created the first 32-bit CD-ROM video game architecture.[26] Launched in 1993, this technology made video games much more realistic and interactive. Software developers could now create games with high-quality video, CD-quality sound, and impressive computer-generated graphics.

3DO's strategy was to make money by licensing software houses to produce games and collecting a $3-only royalty fee — hence the company name. To bring in companies to manufacture the hardware, 3DO licensed its hardware technology for free.

3DO was founded by Trip Hawkins. As a Harvard undergraduate, he designed his own major in strategy and game theory. In 1982 Hawkins, then twenty-eight and a veteran of Apple Computer, had his first big success founding the software house Electronic Arts. The company became known for its computer and video game titles. In 1991 Hawkins gave up the reins at Electronic Arts so that he could devote himself to his new company, 3DO.

Hawkins saw a huge potential market in interactive home entertainment. He observed that Americans spent $5 billion a year going to the movies compared with $14 billion on home video rentals and sales. That arithmetic suggested that the $7 billion a year people spent playing arcade video games should translate to a $20-billion market for home video games. In fact, the home video game market was only $3 billion. Hawkins's plan was to correct the imbalance and capture the missing $17 billion.[27]

In May 1993, 3DO went public at $15 a share, and by October the stock price was $48 — putting the company's value at nearly $1 billion. Hawkins was off to a great start. Investors were obviously enthusiastic, but would consumers share their enthusiasm?

In October 1993 the first 3DO machine came to market. Manufactured by Matsushita and sold under the Panasonic brand for $700, the machine came with *Crash 'N Bum*, a high-speed 3-D racing game. Few other 3DO games were available, and those that were seemed expensive, around $75 a title. By January 1994 Matsushita had sold only thirty thousand machines. These disappointing sales caused 3DO's stock price to retreat to the low $20s.

The economics of CD-ROM-based software made it hard to get the ball rolling. Development costs for CD-ROMs ran up to $2 million per title — due to the sophisticated graphics and audio — compared with half a million for 16-bit game cartridges. Manufacturing costs for CD-ROMs were considerably lower, roughly $2–$4 per disk, compared with nearly $10 for game cartridges. The problem with this economics is that it requires a mass market. A base of thirty thousand machines didn't exactly constitute a mass market, and so software houses weren't willing to invest in writing games for the 3DO machine.

Hawkins realized that he had to crack the chicken-and-egg problem. First, hardware. It was too expensive. Recognizing that it wasn't enough to invite people to play in the hardware game, Hawkins decided to *pay* them to play. In March 3DO offered hardware makers two shares of 3DO stock for each machine sold at a low suggested retail price. Matsushita responded by lowering the price of the Multiplayer to $500, and Toshiba, GoldStar, and Samsung declared their intention to manufacture 3DO machines.

Next, software. There simply weren't enough games. Hawkins's old company, Electronic Arts, had made a big commitment, with twenty-five titles under development. But Hawkins decided that 3DO needed to get in on the software game itself. As product manager Amy Guggenheim explained:

> Looking back, it was not our original intention to do our own title development. We wanted to be a licensing company. But after we launched the system, it became clear that the one company that had 3DO at stake was 3DO. It quickly made sense for us to do software development. By putting some leading-edge titles out in the market place, we benefit everyone. We benefit customers, we benefit software licensees, we benefit hardware licensees. For now, we really have to control our own destiny.[28]

Then back to hardware. It was still too expensive. In October 1994 Hawkins told the 3DO hardware manufacturers that they would have to absorb losses of $200 million over the next fifteen months in order to sell the 3DO machine at competitive prices. When they balked at the prospect, Hawkins turned to the software developers and unilaterally renegotiated the software royalty. Henceforth, 3DO would require software developers to pay a $3 surcharge on every disk sold – above and beyond the $3 royalty they were already paying. A 'Market Development Fund' would be started with this money. Some of the funds would be used to advertise and promote the 3DO machine, while the remainder would be given to hardware manufacturers as an inducement to more aggressive pricing. Hawkins observed: 'If I don't have software companies, that's one problem. But that's not as big a problem as if I didn't have hardware manufacturers.'[29]

By mid-1995 the price of the 3DO machine was down to $400 (with $150 worth of software thrown in). Cumulative sales passed half a million. Progress, surely, but as of early 1996, 3DO's future remains uncertain. It no longer has the 32-bit game to itself. Sega is shipping its 32-bit Saturn machine at $400. Sony has launched its 32-bit PlayStation at $300. Looking to leapfrog them all is Nintendo, whose 64-bit Ultra machine is due out in April 1996 at a price under $250.

The flaw in Hawkins's strategy was relying too much on other players who didn't share 3DO's incentives. That led to problems on both the hardware and the software fronts. 3DO's hardware was never cheap enough. In 1986 Nintendo introduced its 8-bit video game machine at a startling $100, an amount widely believed to be below cost. In 1992 Nintendo and Sega were selling 16-bit video game machines for $100. Why weren't 3DO's machines selling for $100 in 1995? Nintendo and Sega were able to sell the hardware for $100 because they could look forward to making money through future software sales. Matsushita and the other hardware manufacturers had no way to make the money back later.[30] That's why they weren't willing to go down to $100.

The mistake was separating ownership of the software income stream from hardware manufacturing. Hawkins started correcting the problem by handing out equity for each machine sold, but that was late in the game. His second tack, the Market Development Fund, smacks a bit of robbing Peter to pay Paul – only, Peter hasn't been paid yet.

Likewise, there wasn't enough 3DO software, especially early on. Games had to be developed in anticipation of machines that hadn't yet been sold. Others didn't have as much incentive as 3DO to take that gamble. Hawkins started correcting the software problem by developing games in-house, but, again, that was late in the game.

Paying Yourself a Complement The lesson of the 3DO story is: don't rely on others. To develop two complementary markets hand in hand, it's usually better to do it yourself. That's why Nintendo and Sega developed both hardware and software themselves. That's why Intel created ProShare. And that's why 3DO made its life doubly difficult when it tried to leave development of hardware and software to others.

Sometimes people object to the idea of entering a complements business themselves. They say: 'We won't be able to make money there.' That misses the point. You can't look at two complementary businesses separately and insist that they each make some 'target' rate of return. If the accounting rules say you're falling short in the complements market, that's okay. Just think of it as your right hand paying your left hand to play the complements game. The only question that really matters is whether you will make more money overall.

People sometimes object to playing the complements game on another account. They say: 'It's not our business. We should stick to our knitting.' But it's no good clicking your knitting needles if there's no demand for sweaters. You'd better get out of your rocking chair and prod the market.

BRINGING IN COMPLEMENTORS

1. Form a buying coalition on behalf of your customers.
2. Pay complementors to play.
3. Do it yourself: become your own complementor – don't rely on others to develop and price complements aggressively.

Bringing in competitors?

If you don't have a really tough competitor, you ought to invent one ... Competition is a way of life.
— Bill Smithburg, CEO, Quaker Oats[31]

Bill Smithburg has no shortage of real competitors. In sports drinks, his product Gatorade is taking on Coke and Pepsi. In the ready-to-drink iced tea market, his newly acquired Snapple is up against Coke's Fruitopia and Pepsi's Lipton. Smithburg's point is that competition can push you to achieve your personal best. Most runners prefer to train by racing against a rival than by running against the clock.[32]

Management guru Tom Peters tells the story of Quad/Graphics, a company that puts this philosophy into practice: '[It] licenses its most advanced technology to arch-competitors for the express purpose of keeping the heat turned up under itself (and making a buck in the process).'[33]

Of course, you don't want to overdo it. Ideally, you want to get the benefits of competition without giving away the store. Large companies have the luxury of being able to create competition all under their own roofs. Procter & Gamble is famous for encouraging its separate brand managers to compete against one another. Head & Shoulders, Pantene, Pert, Prell, and Vidal Sassoon shampoos are all P&G brands in the same market. Likewise for Tide and Bold detergents and for Ivory and Safeguard soaps. Yet each brand is managed as a separate business with independent advertising, pricing, and strategy. The internal competition across brands keeps everyone on their toes.

Even if you don't think that you need the benefits of competition, your customers may take a different view. Sometimes they won't be willing to do business with you unless you have a competitor, in which case you'd better bring one in.

Chip off the Old Block After Intel developed the 8086 microprocessor in 1978, it provided a second-sourcing license to IBM, Advanced Micro Devices (AMD), and ten other, foreign manufacturers, such as NEC. It effectively gave up its monopoly on this technology. Why did it do that?

Its primary customer, IBM, was quite concerned about investing in developing hardware that relied on the Intel chip and then finding itself with only one supplier. One issue was the question of Intel's manufacturing reliability. At that time, Intel didn't have the track record it has today. IBM insisted on the right to license Intel's microcode in order to manufacture chips for its internal use.

Another issue was the price that Intel would charge in the future. Buyers were worried about the game tomorrow as well as the game today. They required that Intel license second sources. Even though IBM was protected by its right to self-manufacture, it wanted 'backup' protection. By agreeing to widely license its microcode, Intel assured hardware makers that there would be a competitive market for the chip and they wouldn't end up being held hostage. With this guarantee, buyers were willing to commit to Intel's technology. The market for the 8086 chip was indeed very competitive; by 1987 Intel had less than 30 percent of the market.

However, buyers may not have recognized that once they went down Intel's road, it would be difficult to turn back. Intel's commitment to widely license its chip technology did not extend to the 286, 386, 486, Pentium, or Pentium Pro chips. Only five companies were granted a second-source license for the 286. And as things turned out, only IBM ended up with a license for the 386 and beyond, and even that license was restricted to production for internal use.

Why didn't IBM and others look forward and insist that Intel contract to license out every generation of chip technology? In fact, Intel did write long-term licensing contracts with AMD and IBM. But long-term contracts are hard to write, especially in the context of rapidly changing technology. Not surprisingly, legal disputes arose over the interpretation of both the AMD and IBM contracts. AMD lost its second-source license for the 386 and beyond. In 1994, IBM resolved its dispute by selling its licensing right back to Intel for an undisclosed amount. For the Pentium and onward, Intel will be under no obligation to share its technology.

IBM reckoned that if Intel would no longer have to create its own competition, it would have to create some competition for Intel. In a partnership with Apple and Motorola, IBM created its own competitor chip to Intel, the Power PC chip.

BRINGING IN COMPETITORS

1. License your technology both to make money and to avoid complacency.
2. Create second sources to encourage buyers to adopt your technology.
3. Do it yourself: promote internal competition across teams.

That's enough on bringing in more competitors. Most of the time, companies don't complain that they face too few competitors; they think they face too many. Sometimes they even look for ways to have a little less company.

Derailing the Competition Historian Stephen Goddard tells the story of how cars replaced trolleys.[34] In the early days of the car, the 1920s and '30s, Detroit was having trouble breaking into the urban market. In suburban and rural areas, cars and buses had caught on, but downtown the electric trolley was an all-too-effective competitor. It was cheap and convenient. Even worse, trolleys literally stood in the way of the automobile, their tracks taking up the middle of the road.

General Motors, together with Firestone, Mack Trucks, Phillips Petroleum, and Standard Oil, decided to take control. They turned to one Roy Fitzgerald and his four brothers to set up the Range Rapid Transit Company. The mission of this company was to go around the country, buy up local trolley franchises, and shut them down. It was a little more subtle than that, but not much. In towns ranging from Montgomery, Alabama, to Los Angeles, Fitzgerald bought the franchise, took down the electric transmission lines, removed the trolley track, repaved the streets, and junked the electric cars. The deal with the city was that Fitzgerald would replace trolley service with buses. Of course, this was only an intermediate step in moving people toward cars.

Although effective, the strategy was also highly illegal. It was a clear violation of the Sherman Act. The courts fined GM and its co-conspirators $5,000 each; for his role, Roy Fitzgerald was fined $1.

So a mere slap on the wrist for GM and its fellow conspirators. Today the US courts take a much more serious view of anti-trust violations. But there are some circumstances – a declining industry, for example – in which it's both appropriate and legal to acquire competitors in order to rationalize industry capacity.[35]

When he took over as CEO of defense contractor General Dynamics in 1991, William Anders saw his mission as finding a way to rationalize the company's businesses in light of the defense cut-backs following the end of the cold war. As a former US Air Force fighter pilot and Apollo 8 astronaut, Anders held the company's F-16 fighter jet program near and dear to his heart. But with Lockheed also in the fighter business, there was one player too many. Anders tried on more than one occasion to buy Lockheed's F-22 fighter jet program and bring it under his wing. But Lockheed wouldn't sell.

Anders realized that if he couldn't buy out his competitor, he might have to let his competitor buy him out. One of them had to go. If Lockheed wouldn't sell, perhaps it would buy. Although Anders would have preferred to have General Dynamics acquire Lockheed's fighter program than vice versa, he bit the bullet and sold the F-16 business to Lockheed.

3 Changing the Players

Before you enter a game, assess your added value. If you have a high added value, you'll make money in the game; so go ahead and play. But if you don't have much added value, you won't be able to make much money in the game. What then?

Even if you can't make money in a game, you may still be able to make money by changing the game. Ask Cicero's question: *Cui bono?* Figure out who stands to gain from your entry. Those players may be willing to pay you to play.

Holland Sweetener had no added value, but its entry lowered the added value of NutraSweet, and that helped Coke and Pepsi. Holland should have demanded to be paid for up front performing this service. Norfolk Southern had no added value, but its entry lowered the added value of CSX, and that helped Gainesville. Norfolk Southern shouldn't have played without having got paid. BellSouth had little added value in the takeover game for LIN, but its presence lowered the added value of McCaw, and that was to LIN's benefit. BellSouth understood the game and made sure it got paid – handsomely – to play.

If you can't get paid, are you sure you should play? Sitting on the sidelines may be the best strategy. There are more costs to playing a game than are often recognized. Don't ignore the Eight Hidden Costs of Bidding.

Once you're in a game, you can try to change who else is in the game. Go around the Value Net and consider bringing in customers, suppliers, complementors, even competitors. Remember that these other players, just like you, are free to choose whether to play. You can't force them to play, so you have to create the right incentives.

Sometimes that's possible and very effective. American Express's strategy of forming a buying coalition is a perfect example: it attracted many more bidders into the game by giving everyone a greater incentive to seek its business. With more bidders in the game, each had less added value, and that put the buying coalition in a stronger bargaining position. There are some roles, however, that you shouldn't leave to others. An example is developing a complement that's critical if you're to have any added value. When the stakes are this high, don't count on others. The best person to play the part may be you. That was the lesson of 3DO.

Anytime the cast of players changes, so do added values. Added values can also be changed more directly, as we'll see in the next chapter.

5 Added Values

> **Nothing is more useful than water; but it will purchase
> scarce anything; scarce anything can be had in exchange for
> it. A diamond, on the contrary, has scarce any value in use;
> but a very great quantity of other goods may frequently be
> had in exchange for it.**
> — Adam Smith, *Wealth of Nations*, 1776

Two hundred years ago, Adam Smith presented a paradox concerning water and diamonds: Water is essential to life, while diamonds are not. Yet water is essentially free, while diamonds, alas, are not.

Had Adam Smith lived in the 1990s, he might have compared televisions and cars with video games. Televisions and cars are like water, practically essential to our modern lives. Video games are modern diamonds. They're a kid's best friend, but no more. So which business would you prefer to be in: consumer electronics, cars, or video games?

Let's compare three prominent Japanese companies, one in each business: Sony, Nissan, and Nintendo. Sony makes televisions; Nissan makes cars; Nintendo makes video games, especially one about a plumber named Mario. Between July 1990 and June 1991, the average market values of these three companies were:

Nissan	2.0 trillion yen
Sony	2.2 trillion yen
Nintendo	2.4 trillion yen

Yes, for a while at least, Nintendo was worth more than either Sony or Nissan.[1] How did Nintendo do it? Never mind how it *surpassed* Sony and Nissan; how did it even come close? The answer lies in added value.

1 Added Value of a Monopoly

One strong company and the rest weak.
– Hiroshi Yamauchi, President, Nintendo[2]

Video games date back to 1972 with the founding of Atari Corporation.[3] Atari, although an American company, took its name from the Japanese game of Go. The meaning is akin to 'check' in chess – the announcement that a rival's territory is under attack. Having given fair warning, Atari proceeded to capture the market with a table tennis video game called *Pong*.

Atari's success was dramatic but short-lived. In just ten years, the US home video game market grew from nothing to $3 billion in retail sales. But the market was inundated with poor-quality software, and that led to its demise. In 1985, sales fell below $100 million; home video games were dismissed as a fad. Atari lost as much money on the way down as it had made on the way up. People wrote the industry off.

So no one was paying too much attention when Nintendo came on the scene. A century-old Japanese company, Nintendo had, over the years, gradually expanded from making playing cards to making toys, and then to making arcade games. Outside Japan, Nintendo was virtually unknown. But not for long.

Nintendo Power Loosely translated, Nintendo means: 'Work hard, but in the end it is in heaven's hands.'[4] Actually, with its move into home video games, Nintendo didn't leave much to chance. It did everything right. Nintendo set in motion a virtuous circle.

First and foremost, the hardware was a bargain. Nintendo had found a way to reproduce the feel of an arcade game on an inexpensive home machine. The result was a new video game system called the Famicom (Family Computer). Nintendo launched the Famicom in Japan in 1983 and brought the machine, renamed the Nintendo Entertainment System, to the United States in 1986.

In truth, the Famicom was hardly a computer at all – everything was dedicated to a single purpose, game playing. In order to keep the costs down, Nintendo deliberately used a commodity chip, an

8-bit microprocessor dating back to the 1970s. Personal computers at that time — such as the IBM AT or the original Apple Macintosh — were selling for between $2,500 and $4,000. Nintendo's machine was priced at 24,000 yen (around $100). The Famicom's price radically undercut the competition, its price so low that many people believed it to be below cost.

Along with its bargain hardware, Nintendo also had superb games. This was no accident. Nintendo used its experience in arcade games to develop a new level of home video game excitement. Its ace designer, Sigeru Miyamoto, was a genius at pushing the performance envelope, creating such smash hits as *Donkey Kong, Super Mario Bros.*, and *The Legend of Zelda*.

With the inexpensive hardware and a selection of hit games, consumers began buying Nintendo's machines and games in large numbers. The home video game industry was back in business.

The Virtuous Circle Once sales took off, Nintendo didn't have to do everything itself. Software houses lined up to write games for the Nintendo system, but they couldn't without Nintendo's permission. Wary of the previous industry crash, Nintendo had built a security chip into the hardware to ensure that only Nintendo-approved cartridges could run on the system. The idea was to prevent the rampant copying that had previously destroyed the industry. Software houses could write for the Nintendo system only if Nintendo let them. The result? Nintendo had complete control.

Nintendo's software licensing program had some remarkable conditions. Each licensee was limited to just five titles a year. That way, developers had to emphasize quality over quantity. All games had to meet a set of standards that included a ban on any excessively violent or sexually suggestive material. Nintendo did all the manufacturing of the approved games. It made its money by charging licensees a large markup on every game cartridge. On top of all this, an exclusivity clause prohibited licensees from releasing the same title for other video game systems for two years.[5]

The result was a virtuous circle. The cheap hardware and Nintendo's own hit games got it started. As more consumers started buying the hardware, Nintendo could drive down its manufacturing cost. With a growing base of machines, Nintendo was able to attract outside game developers. This created a positive feedback loop. With more and better games, still more consumers bought Nintendos, leading to a larger base, still lower costs, and even more games. With even more games, Nintendo hardware became even more valuable, leading to yet further sales. The upshot was 'Nintendomania.'

Even as demand took off, Nintendo remained cautious about flooding the market. It strictly controlled how many copies of games were produced, and pulled its own games off the market as soon as interest declined. Over half of Nintendo's game library was inactive. Sometimes, severe shortages resulted. The most notable was in 1988, when retailers requested 110 million cartridges, and perhaps as many as 45 million units could have been sold. But only 33 million units were available.[6] The shortfall became especially acute during the Christmas season.

Somewhat paradoxically, the shortage may have helped create even more consumer demand. There were at least three different effects going on. First, shortages made the game cartridges even more desirable in the eyes of consumers, actually boosting demand. Trendy restaurants play the same game. For example, the long lines outside K-Paul's in New Orleans made it even more fashionable, further increasing the lines.[7] Why not just raise the price and eliminate the queue? Because eliminating the queue might reduce the product's cachet, and demand could crash.

Second, shortages made headlines; filling demand would not have. 'Tonight's top story: Nintendo sold game cartridges to all those who wanted them. Details at eleven.' We don't think so. The shortages generated tremendous free publicity for Nintendo, a company known to be rather stingy on advertising (spending only 2 percent of sales).

Third, shortages helped retailers move slower-selling Nintendo games, because parents would buy a lower-selling title if the one their kid wanted was sold out. Of course, this was only a temporary solution, what we call the Band-Aid effect. The substitute might tide the kid over from Christmas to New Year's, but kids tend to remember these sorts of things. So parents would have to return for the sold-out title once fresh supplies came in. Nintendo made two sales instead of one.

It was truly a case of 'less is more.'

To further stimulate interest in the games, Nintendo created a monthly magazine that rated games, gave playing tips, and previewed upcoming games. There were no ads. Nintendo kept the magazine as cheap as possible – it was priced simply to break even. By 1990 *Nintendo Power*, with an estimated 6 million readers, had the highest circulation of all US children's magazines. The magazine was an ideal complement to Nintendo's games.

The result: By the end of the decade, Nintendo had rebuilt home video games to a $5-billion worldwide business. It had achieved an incredible 90-plus percent share of the Japanese and US 8-bit video game markets. A Nintendo could be found in one out of every three Japanese and American households. Nearly three-quarters of US households with teenage boys had video game systems. Nintendo's products accounted for over 20 percent of the entire US toy industry, with cumulative sales of the Super Mario Bros. game series, alone, topping 40 million copies. Mario was more popular than Mickey Mouse among US children.[8] More popular than Mickey Mouse? Yes.

Could a challenger hope to breach Nintendo's virtuous circle? Not once the circle had got rolling. Forget about alternatives – TV, books, sports. From a kid's perspective, there were no good alternatives to a video game. The only real threat came from alternative video game systems. Here, software was key, as always. With a huge library of Nintendo titles to choose from, why would anyone buy another machine? Perhaps a challenger could take successful Nintendo games over to its platform and then offer its own library. But the exclusivity clause killed that option. No game could be taken to another platform for a two-year period, by which time the game was passé. A challenger would have had to start from scratch. While large profits and shortages normally invite entry, the virtuous circle made competing in Nintendo's game hopeless. The only hope was to leapfrog Nintendo with a new technology; that's what Sega ultimately did, as we'll see in the Scope chapter.

Power Play Because Nintendo had a monopoly in 8-bit video game machines, its added value equaled the entire home video game pie. There was no threat from competitors. No other hardware maker, whether an Atari or someone else, had any added value.

But there were other players who had claims on the pie. Go

around the Value Net. There was the retail channel – Nintendo's proximate customers – which included megaretailers such as Toys 'R' Us and Wal-Mart. There were Nintendo's complementors, outside game developers such as Acclaim and Electronic Arts. And there were Nintendo's suppliers. These included chip manufacturers, such as Ricoh and Sharp, and the owners of cartoon characters and comic-book heroes – such as Disney (Mickey Mouse) and Marvel (Spider-man) – used in some Nintendo games.

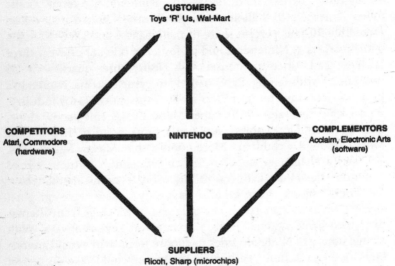

Nintendo's strategy had the effect of limiting the added value of each of these other players in the game.

Start with Nintendo's customers. How did Nintendo combat the buyer power of a Toys 'R' Us or Wal-Mart? Nintendo's inventory management policy was the key. Cartridges were constantly in short supply. Nintendo may have lost some sales, but the more important effect was that some retailers couldn't get supplied. The retailers' position was similar to that of the students in Barry's version of the Card Game. It was as if there were twenty-six customers – actually, there were many more – chasing after Nintendo's twenty-three black cards. Just like Barry's students, the retailers had little, if any, added value. Even a giant like Toys 'R' Us was in a weak position. As

Nintendomania took hold, consumers queued outside stores, and retailers clamored for more product. With games in short supply, Nintendo had zapped the buyers' power.

The next arena of negotiations concerned the software developers. What happened here? First, of course, Nintendo was in this business itself, doing quite well. Then the security chip allowed Nintendo to set up a carefully managed licensing program. The restriction to five titles a year kept the developers symmetric; no one developer could become too powerful. And because Nintendo developed games in-house, it was even less dependent on any one outside developer. The combined effect was to sharply limit the added values of the licensees.

Nintendo's suppliers, too, had little added value. The chips were a commodity. As for the game characters, Nintendo hit the jackpot by developing Mario. After Mario became a star, the added values of Mickey Mouse, Spiderman, and other licensed characters were reduced. In fact, Nintendo turned the tables completely, licensing Mario to appear in comic books and on cartoon shows, cereal boxes, board games, and toys.

The name of the monthly magazine, Nintendo Power, summed up the situation quite nicely.

Nintendo's strategy resulted in a high added value for itself and a low added value for everyone else. In this way, Nintendo was able to capture a giant slice of a medium-sized pie. By contrast, Sony and Nissan are in competitive businesses – each faces many rivals. Don't confuse the added value of Sony with the much larger added value of televisions, or the added value of Nissan with that of cars. You can still watch television without Sony, although not on a Watchman or a Trinitron, and you can still drive without a Nissan, although not in a Maxima or an Infiniti. As for Nintendo, because it had a monopoly, its added value was essentially the added value of video games. Without Nintendo, there would have been no video-game game. That's the key to how Nintendo was able to surpass the market values of Sony and Nissan.

The Antitrust Challenge Amidst a climate of strained US-Japanese relations, some people called Nintendo's practices into question. In late 1989 – Pearl Harbor Day, in fact – Congressman Dennis Eckart

(D-Ohio), chairman of the House Subcommittee on Antitrust, called a press conference to ask the Justice Department to investigate allegations that Nintendo unfairly reduced competition.

Eckart's letter to the Justice Department outlined several areas of concern. The first was the alleged anticompetitive purpose of the security chip. The second turned on the licensing agreements with software developers who, the letter alleged, 'became almost entirely dependent on Nintendo's acceptance of their games and production allocations.' The letter further alleged that the Christmas shortages in 1988 were 'contrived to increase consumers' price and demand and to enhance Nintendo's market leverage.' Finally, the subcommittee charged that Nintendo had 'aggressively exercised its market power' by threatening retailers with limits on their supply, or even with cutting them off, if competitors were given shelf space. Eckart asserted that because of these practices, 'the net result is that there is only one game in town.'[9]

Commenting on the government's antitrust case, the business weekly *Barron's* wrote:

> **The legion of trustbusting lawyers would be far more productively occupied playing Super Mario Brothers 3 than bringing cases of this kind ... In their pursuit of ... crooks, we wish the trustbusters well. But [they] are in equally hot pursuit of Nintendo and other real business success stories, real achievements, real technological progress and real rewards.**[10]

A year later, the government dropped its investigation of Nintendo.[11]

Monopoly money

> **Always leave them asking for more.**
> – Old vaudeville saying

What does it mean to have a monopoly? Without you, there's no game. So your added value is equal to the whole pie – an enviable position to be in. But how well you do depends not only on your added value but also on the added values of everyone else making claims on the pie. That's where shortages come into play.

In the Card Game, both Adam and Barry had a monopoly. Each had all the black cards, but Barry did a lot better than Adam. By losing three cards, Barry effectively created a shortage. The result was that Barry's students had a lot less added value than did Adam's. So while Adam split the pie evenly with his students, Barry captured the lion's share. That's why Nintendo did so well: while monopoly by itself is nice, monopoly and shortage is twice as nice.

The Nintendo story demonstrates the potent effects of under-supplying customers. This strategy has a counterpart toward suppliers, as must be the case, given the symmetry of the Value Net. The counterpart strategy is underdemanding a resource – a strategy that is potentially as powerful as undersupply.

The National Football League employs both strategies. In the Game Theory chapter, we saw how the National Football League limits the supply of teams. The league also limits its demand for players. By holding down the number of teams and limiting the roster size of each one, the NFL ensures that there are many more players who would like to play professional football than will ever get the chance. That reduces the added value of every football player, even someone who makes the cut. Were the NFL to expand, the quality of the play might go down, but the players' pay would go up.

The Ace of Diamonds Let's go back to where we began this chapter: Adam Smith's water-diamond paradox. Two hundred years ago, diamonds were to be found only in the riverbeds of India and the jungles of Brazil.[12] Thus, supply and demand explained the relative prices of water and diamonds. Water was plentiful and therefore cheap. Diamonds were scarce and therefore dear. Paradox lost? Not quite.

The real puzzle is why diamonds have remained so expensive. The high prices led people to search for new sources – and they found them. In the 1870s major diamond mines were discovered in the Transvaal of South Africa. Still more diamond reserves have been found in Angola, Australia, Botswana, Namibia, and Zaire. In the 1960s Russia found a way to unearth its massive reserves from under the Siberian permafrost. It's now the world's largest producer of high-quality gemstones. Between 1950 and 1985 the world's output of diamonds grew from 15 million to 40 million carats a year. Then, between 1985 and 1996, it more than doubled again to over 100 million carats. Today diamonds are hardly scarce.

That's one side of the equation. As for demand, it's much more stable. Demand is mostly driven by demographics – the number of people getting engaged, in particular. With an increasing supply and relatively stable demand, why is the price still so high? It seems the water-diamond paradox is more paradoxical today than it was in Adam Smith's time.

There's a one-word explanation: DeBeers. This South African company has a monopoly on the world's diamond market. Nearly all of the world's diamonds are sold through DeBeers's distribution system, known as the Central Selling Organization. Even Russia is committed to selling 95 percent of its diamonds through DeBeers.

There's more. DeBeers holds back the supply. Recall how Nintendo didn't fill all of its retailers' orders in 1988. With DeBeers, every year is that way. It holds ten sales a year – 'sights' in industry parlance – to which only a select 150 diamond merchants are invited. DeBeers decides on the appropriate supply. Each invitee then gets an allotment of stones in a plain brown shoe box. It's take it or leave it, and the dealers usually take it. Dealers who try to circumvent DeBeers – by hoarding, speculating, or dealing with the black market – shouldn't count on being invited back to the next sight. There's no shortage of dealers who would be glad to take their place.[13]

As well as controlling supply, DeBeers has made sure to manage demand. People value diamonds highly because they perceive them to be scarce. They're not, but that's beside the point; it's the perception of scarcity that counts. The perceived scarcity has made diamonds an ideal choice for engagement rings, but not without some help from a long-running DeBeers advertising campaign. In fact, the diamond engagement ring is a newly created 'tradition' in Japan. In 1967 only one in twenty Japanese brides had a diamond ring. Now it's almost compulsory.

DeBeers uses advertising to shape the game in other ways, too. It wants people who buy diamonds to keep them forever. This limits competition from the secondhand market. So, to discourage people from selling their diamonds, DeBeers created its 'Diamonds Are Forever' campaign. In response to an unwelcome upsurge in Russian exports of medium-grade diamonds, DeBeers popularized the 'eternity ring,' an anniversary ring studded with these medium-grade stones.

But despite all its efforts, DeBeers has been unable to keep demand ahead of the rapidly growing supply. So, to maintain the scarcity of diamonds, it's stockpiling more and more stones. While DeBeers sold a record $2.5 billion worth of diamonds in the first half of 1994, by June of that year, its inventories reached $4 billion – double the levels of a decade earlier.

The Russian Treasury has its own growing hoard of diamonds, estimated to be worth at least another $4 billion.[14] With the recent political instability in Russia, this inventory makes a very tempting cookie jar, and unauthorized stones have been getting onto the market. In response, DeBeers has had to further cut back its own supply just to keep prices flat. Not for the first time in its history, DeBeers faces a serious challenge.

DeBeers and the Russians have a mutual interest in regaining control of supply. This time they may succeed. But it's getting harder and harder to maintain a scarcity of diamonds in the face of the growing natural abundance. Diamonds may be forever, but how much longer will Adam Smith's paradox persist?

The explanation of the modern water-diamond paradox goes beyond supply and demand. It's about one player owning the supply and being able to control it. To see the significance of ownership and control, imagine for a moment that all the water in the world is owned and controlled by one person, a DeBeers of water. Next to this, the DeBeers of diamonds would look like a mere drop in the ocean.

Limiting supply?

Let's apply the lesson of the Nintendo, NFL, and DeBeers stories to a more familiar business situation. You face growing demand for your product. It's clear that more capacity is needed, but how much more? Management science textbooks frame this as a balancing problem: expand too little and you risk losing sales; expand too much and you'll end up having paid for unused capacity. The optimal expansion minimizes the expected costs of these two mistakes.

The textbook analysis of capacity expansion assumes that the profit from a sale is the same, regardless of whether there's a shortage or a surplus. That's a bad assumption. There's a big asymmetry between the two cases. Underbuild a little, and each customer has little added value. Overbuild a little, and every customer is as powerful as you are. Your margins will be much healthier when there's a shortage rather than a surplus. It's the Card Game once again. Just a small change, up or down, in the number of black cards made a big difference in how Barry fared relative to the students.

Thinking in terms of added value tips the balance in favor of building less rather than more. You'd rather err by having too little capacity. Excess capacity carries a big cost – in terms of lost bargaining power – that is often overlooked.

The surprising asymmetry between shortages and surpluses helps explain the recurring cycles seen in industries ranging from pulp and paper to chemicals to hotels to memory chips to property and casualty insurance. Someone expands capacity and suddenly everyone's profits fall precipitously. Even a very little excess capacity can cause profits to fall a very long way. The sharp fall in profits leads to a moratorium on capacity expansion.[15] Meanwhile, demand continues to grow, and before too long, there's insufficient capacity. Power shifts back to the producers, and now they're back in the black. Someone gets enthusiastic and overexpands, even just a little, and the cycle repeats itself.

So far, the message is: beware oversupply. But don't go overboard. There are some hidden costs of undersupply that you need to factor into the equation. Undersupplying products shrinks the pie: it leaves unsatisfied demand today. And a lost sale today can mean a lost relationship – lost future sales, too. The shortage is also likely to create ill will. Customers who don't get the product will certainly be unhappy, and even those who do get supplied may resent the high price you charge. There's an offsetting factor if the shortage creates a cachet effect – as with Nintendos and diamonds – but that shouldn't be taken for granted.

In short, undersupply creates a hole in the market and disenchants customers. It's an invitation for others to enter. Even the customers you're selling to might welcome the chance to switch and teach you a lesson. That's why you might do better in the long run by playing Adam's version of the Card Game rather than Barry's. You'd sacrifice some profits today, but you'd keep the game going.

LIMITING SUPPLY

PROS

1. Gets you a bigger slice of the pie.
2. May give you cachet.
3. May provide free publicity.
4. May lead customers to buy your slower-moving products while waiting for the shortage to end.

CONS

1. Shrinks the pie – costs you sales today.
2. May cost you a relationship and thereby future sales.
3. Creates ill will.
4. Leaves a hole in the market inviting entry.

So far we've been assuming that you have some added value. The focus was on whether and how to limit the added values of the other players in the game. Of course, you can't take your added value for granted. The next section takes a look at ways to engineer added value.

2 Added Value in a Competitive World

In a competitive world, you have to work hard to have any added value. This hard work is what a lot of basic business is about. You find ways to make a better product, and you look to use resources more efficiently. You listen to your customers to learn how to make your products more attractive to them. You work with your suppliers to discover ways to run your business more efficiently for you and more effectively for them. You step into the shoes of your customers and suppliers and understand their perspectives.

But life isn't that simple. There's a catch: improving the product increases the cost. Likewise, if you cut costs, you compromise your product. There's a quality-cost trade-off. You can have higher quality or lower costs, but not both.

Trade-offs

One way to engineer added value is to make intelligent trade-offs. The trick is to spend $1 in such a way that customers value the quality improvement at $2. Then you can raise price by $1.50, and it's a win-win. Likewise, the trick is to save $2 in such a way that customers value your product only $1 less than before. That way, you can cut price by $1.50, and it's a win-win. In both cases, you've engineered an extra dollar of added value and split it with the customer.

To find these trade-offs, you have to depart from business as usual. You have to challenge the old, comfortable assumptions about how you operate – or, in TWA's case, the old, uncomfortable assumptions.

Comfy and Cozzi In January 1993 TWA was in Chapter 11 reorganization. The airline was in a nosedive. Passengers were abandoning it. TWA was at the bottom of consumer ratings – by an uncomfortable margin. Employee morale was nonexistent. And there was only $10 million left in the coffers.

Bob Cozzi, TWA's senior vice-president of marketing, saw a way out. He proposed removing ten to forty seats per plane and spreading out the remaining seats to give passengers in coach more legroom. According to Cozzi, 'We rolled the dice right down the table. We spent $1 million to take out the seats, the other $9 million to promote it.'[16] It was an all-or-nothing bet.

Cozzi promoted Comfort Class with the ad campaign 'TWA – the most comfortable way to fly.' Comfort Class offered an extra three inches of legroom, a big difference when the industry standard was thirty to thirty-two inches. While everyone offered more legroom in business and first class, TWA was the only major airline to add legroom in coach. Still, many people were skeptical. Unkind observers likened the initiative to rearranging the deck chairs on the Titanic.[17]

The skeptics were proved wrong. Customer satisfaction soared, as did employee morale. Within six months, TWA moved from the bottom to the top of the rankings – all because of the extra legroom. TWA was rated below average in six of seven categories: on-time performance, aircraft interior, flight accommodations, scheduling, in-flight amenities, gate check-in, and postflight performance. Yet its performance in the remaining category, seating comfort, so clearly dominated all the other airlines that market research firm J. D. Power ranked TWA the top carrier for long domestic flights and number two for shorter flights.

All this helped fill up the planes. Cozzi calculated that an extra coach passenger on every flight was worth $80 million per year. Equally important, many more full-fare travelers were now flying on TWA. If a company didn't permit its employees to fly business class, TWA's Comfort Class was the next best thing. By the end of 1993, yield or average revenue per seat was up by 30 percent — double the figure for the rest of the industry.

Comfort Class was a very smart and cost-effective way for TWA to improve its quality of service. The real cost of taking out a seat is the lost revenue from someone who would have paid to sit in that seat. If the planes aren't full, it costs very little to take out some seats and give people more legroom.

This was a win for TWA and a win for its customers. With its improved service, TWA also had a leg up on the competition. Did that mean it was a loss for the other airlines? Not necessarily. To the extent that TWA attracted full-fare passengers from them, yes, it was. But there was an element of win-win present as well. TWA was not about to start a price war. With fewer seats and even fewer empty ones, it had no incentive to cut price. In fact, with customers willing to pay more for its improved service, TWA even had some room to raise price. The other airlines benefited now that TWA was no longer forced to compete on price.

But what if other carriers copied the strategy? Wouldn't that negate TWA's efforts? No. If others copied TWA's move, excess capacity would be retired from an industry plagued by overcapacity. Passengers would get more legroom, and carriers would have fewer empty seats to tempt them into starting price wars. Cozzi had found a way to move the industry away from the self-defeating price competition that goes on when airlines try to fill up the coach cabin. This was business strategy at its best.

As it turned out, most other carriers didn't follow TWA's initiative. They were concerned about the 'Friday afternoon effect.' At peak travel times, especially Friday afternoons, planes fill up. On those occasions, every removed seat has a real cost. The other airlines decided to keep their seats, even if the seats remained empty during the rest of the week, so that they would be available for peak usage. What the airlines missed is that taking seats out changes the game during the rest of the week. An airline with fuller planes – even if that's because seats were taken out – has less incentive to cut price. The gain from higher prices could be much more important than the few lost Friday afternoon passengers. Apparently, the other airlines didn't see it this way.

In fact, TWA forgot its own logic and almost aborted Comfort Class. The strategy was so successful in filling up planes that giving the extra legroom started looking a lot more costly. In 1994 new management came in and decided that the extra comfort didn't justify the cost, forgetting why the planes were full in the first place. As Cozzi describes it, 'Comfort Class was going so well that the new management decided to kill it.'[18] The employees rose to defend Comfort Class. Faxes protesting the move, over three hundred of them, came flying in.

Cozzi resigned in opposition to management's plans. Finally, management retreated, and in April 1994 Comfort Class was scaled back rather than eliminated. Seats were put back only on planes earmarked to serve markets experiencing strong summer demand. In the fall of 1994 TWA returned to full-scale Comfort Class on its domestic flights. It's not too late for other airlines to introduce their own versions of Comfort Class.

Comfort Class was hardly a trade-off. That was its brilliance. The higher-quality service came at a low incremental cost. More often, quality really costs you. And then you can't assume that everyone will be willing to pay enough to justify the extra cost. Different people place different values on quality. Some people will inevitably value a quality improvement by less than the increase in cost. Others will be willing to pay a lot extra, even for small improvements in quality. You spend $1 to improve quality. Some people might then be willing to pay an extra $10, others an extra $2, and still others only an extra 50 cents. That makes these trade-offs a bit tricky to manage. It's a numbers game.

Everyone agrees that the Concorde supersonic airplane is a better way to fly. However, it has a much higher operating cost, primarily due to its extremely limited capacity. There don't seem to be enough people for whom the three-hour time saving justifies the higher cost. The Concorde has only one hundred seats, and even those are seldom all filled. But that doesn't mean that the idea of supersonic travel isn't economically viable. A larger supersonic plane, with lower costs and hence lower fares, would attract more travelers and could be profitable. Back in the 1960s, when the Concorde was designed, a larger supersonic plane might not have been an engineering possibility; today it is.

The flip side of raising quality is saving costs. As you try to save costs, people may have different reactions. Many won't mind a small reduction in quality. But others will now value your product quite a bit less — for them, the cost savings don't pay off. Again, it's a numbers game.

Taco Bell and British clothing retailer Marks & Spencer get the numbers game right. You can get Mexican food that is better than Taco Bell's and better-quality clothing than Marks & Spencer has to offer. But the higher quality comes at a cost that most people aren't willing to bear. Taco Bell and Marks & Spencer forgo some customers, but gain many more. In fact, Taco Bell and Marks & Spencer offer such good value for money that their customers are eager to make the trade-off.

The examples of Comfort Class, Taco Bell, and Marks & Spencer all involve minimal trade-offs. At a small cost, TWA engineered a big improvement in quality. Likewise, for a small reduction in quality, Taco Bell and Marks & Spencer engineered large cost savings. Ideally, the trade-off should be as small as possible. The least trade-off is to make no trade-off at all.

Trade-ons

While trade-offs are one way to engineer added value, even better is when you end up with higher quality and lower costs at the same time. This outcome is what we call a trade-on. Do trade-ons exist? Absolutely. Think of the quality revolution. People learned that redesigning the manufacturing process — rather than reworking defective items — led to quality improvements and cost savings at the same time. They found that high quality is low cost.

Establishing a virtuous circle is another route to creating trade-ons. You make a better product. At first, your costs rise by more than the perceived quality improvement. But if you can hang on, more customers will come to you. With larger volumes you can operate more efficiently. The quality-cost trade-off begins to work in your favor. Now you're earning more money than at the outset. You can invest some more in product improvement, lower price, or do both. You get still more customers and still greater efficiencies. The virtuous circle is rolling. With enough scale, your costs might even be lower than at the outset. You've turned a trade-off into a trade-on.

In the current debate over protecting the environment, it's often said that there are inescapable trade-offs. It's possible to have cleaner and greener products, but only in return for accepting lower quality or higher costs. Not so. It's possible to have environmental trade-ons, as documented by Harvard Business School professor Michael Porter and his coauthor, Claas van der Linde, a management professor at Switzerland's St Gallen University. They point to the case of the Dutch tulip business.[19] To reduce contamination of soil and groundwater, the Dutch have moved cultivation of tulips from outdoors to advanced greenhouses, where pesticides and fertilizers are now recirculated in a closed system. The controlled environment also reduces infestation risk, allowing growers to further economize on pesticides and fertilizers. Moving indoors, growers have found a new way to plant tulips, which has helped lower handling costs. Finally, the greenhouses reduce variability in growing conditions, improving product quality. The Dutch tulip growers have managed to protect the environment and enhance their added value at the same time.

Discovering trade-ons is not just an engineering job. There are opportunities everywhere, if you go out and look for them. Dedicated authors that we are, we understand the need to do field research – which brings us to Club Med.[20]

Getting into Bed with the Customer While no one's asking for any sympathy, it's hard work being on the staff at Club Med. The day starts with breakfast at 7:15 and often goes well past midnight, playing late-night basketball, bartending, or rehearsing skits for the next evening's performance. Staff members need to speak several languages and work a six-day week. And yet Club Med pays its staff – the *gentils organisateurs* – significantly less than the market wage for multilingual college graduates. Labor costs, as a percent of sales, are around ten points below the hotel industry average.

Does that mean Club Med gets third-rate employees? Not at all. Turnover and growth create 2,000 openings a year, for which there are over 35,000 applicants. Club Med attracts people for whom money isn't a priority. Only those who really love the Club Med experience apply for the job. Club Med even recruits staff from its guest list, resulting in a staff that is a whole lot like the clientele. So everyone gets along.

By economizing on labor costs, Club Med ends up enhancing its appeal to customers. A trade-on.

Club Med also keeps costs down by finding ways to keep guests entertained within the resort's compound. There's so much going on – trapeze lessons, windsurfing, evening performances put on by the *gentils organisateurs* – there's no reason to leave and explore the island. In fact, Club Med doesn't encourage activities outside the compound. Field trips, while available, are rather expensive. Part of the enjoyment of Club Med is spending time with the other guests. If some people leave to explore the island, this diminishes the group experience for the others.

Directing activities inside the compound also helps brand the product. The overriding feeling that people get on a Club Med vacation is one of having gone to Club Med, not of having visited Martinique or Paradise Island. The Club Med experience revolves around Club Med, not the island on which it's located.

The facilities at Club Med are simple. Accommodations are perfectly functional – some might say spartan, no one would say lavish. There are no phones, clocks, televisions, newspapers, not even writing paper. The low-cost quarters are designed to get people out of bed and into the common outdoor spaces. That promotes the group experience. Once again, Club Med reduces its costs and increases its appeal at the same time.

Club Med leases most of its properties. But its strong brand and simple facilities mean that it isn't vulnerable when the lease comes up for renewal. Club Med could always move to a different part of the island without losing its customers or having to write off a large investment. The added value is with Club Med and not the landlord.

Club Med is a case of less is more. Its low-cost strategy leads to happier customers. Keeping things simple has made a better product.

What about creating trade-ons the other way round? Can making customers happier lead to cost savings? Absolutely. Here's the inside story.

Captive Market The Tennessee legislature has concluded that the prisons run by a private enterprise, Corrections Corporation of America (CCA), are both better and cheaper than the ones it runs itself through the state department of corrections.[21]

The fact that it's possible to run a prison for less money comes as no surprise. As Norval Morris, emeritus professor of law and criminology at the University of Chicago, puts it: 'Obviously, you can build a dungeon and throw people in it and throw food down to them very cheaply. The question is what services you provide them.'[22]

The interesting point is that CCA has managed to save costs while at the same time running a better prison. CCA charges Tennessee a daily rate of $35.18 per inmate, while comparable state-run prisons cost $35.76. That might not look like a big cost savings, but bear in mind that CCA makes money at the price it charges. In 1994 CCA's profit margin was 7.3 percent, averaged across the forty-five prisons it runs. As for quality, a special committee of the Tennessee state legislature gave the CCA prison a higher score than comparable state-run prisons. CCA's prisons had fewer escapes and inmate assaults and offered better medical care and more job and education programs.

The view from the inside is equally positive. Phillip Phillips, a twenty-five-year-old serving a ten-year term for armed robbery, judges the CCA prison the best of the six he's been in. 'It's cleaner, you get more choice of food, and the staff is more patient and willing to take time.' 'I should never have left,' laments Samuel Mitchell, a twenty-one-year-old convicted robber who transferred to one of the state-run prisons to be near his brother.[23]

How does CCA do it? The biggest expense in running a prison is the guards' salaries. In a state-run prison, up to 25 percent of the entire budget can go to overtime pay. Doctor Crants, founder and chairman of CCA, figured out why. If a prison is unsafe and depressing, the guards don't want to come to work, and they call in sick. Other guards have to work overtime, and up go labor costs. According to Crants: 'What this is all about is to make the corrections officer think he is coming to work in a nice place. So don't overcrowd the inmates, give them lots of programs to keep them busy and keep the walls painted and the grass green.'[24]

The dungeon approach isn't the cheapest, after all.

Private-enterprise prisons are a growing phenomenon. We suspect that there are many other public-sector activities that could deliver higher quality while at the same time saving costs. Any sector of the economy where market forces have historically been absent is a good place to look for trade-ons.

ENGINEERING ADDED VALUE

TRADE-OFFS

1. Raise the amount customers are willing to pay by more than the incremental cost.
2. Reduce cost without reducing willingness to pay by as much.

TRADE-ONS

1. Lower costs in a way that helps you deliver a better product.
2. Deliver a better product in a way that helps you lower costs.

---

3 Added Value of a Relationship

You do your best at figuring out how to provide high quality at low cost. But so do your competitors. That's the nature of competition. If there are many others who can do what you do, then you don't have much added value.

When you don't have much added value, you can't sustain much of a premium over cost. You don't make much money. It's even worse if your business has low variable costs relative to fixed costs. Then you could well be unable to cover your fixed costs, in which case you end up losing money.

Examples of businesses with significant fixed costs come easily to mind: airlines, car rental agencies, health clubs, hotels, and restaurants. Many commodity businesses – aluminum, chemicals, oil refining, pulp and paper, and lots more – have similar economics. A common denominator is that all these businesses have to run their operations more or less independently of how many customers show up.

That by itself isn't a problem. Some companies in these businesses do quite well. For example, the Four Seasons hotel on Nevis offers a unique Caribbean vacation; it has a high added value. But the downtown hotels – Hilton, Hyatt, Marriott, Sheraton – to be found in almost any city in the world have much less to distinguish themselves from one another. That puts all of them in a weaker position. Each has a large added value when the market is tight. But if the market gets overbuilt or demand slacks off, profits tumble.

As for airlines, they're in a very vulnerable position. Sure, there are some routes on which one airline has a large added value. If you want to fly nonstop into Minneapolis, the odds are that you're going to have to travel on Northwest Airlines. There are hardly any other options. But on most routes, there are several alternative carriers, and the flying experience on one isn't that much different from that on another. Add to this the fact that there is significant excess capacity in the industry and that almost all costs are fixed, and you have a good explanation of why airlines face such a challenge making money.

Facing the biggest challenge, the airlines have been the most creative in coming up with solutions. They've long understood the power of engineering a relationship with customers as a way to engineer added value in a competitive market. That's the point of frequent-flyer programs, as we're about to see.

Creating loyalty

The $64,000 question is how to develop a relationship. To some extent, relationships are automatic. After customers have bought from you once, next time they have a natural incentive to buy from you again, rather than from the competition. It's simple inertia. Familiarity breeds content. That gives you some added value, but it may not be enough.

You can do more. You can actively promote strong relationships with your customers – and with your suppliers, too, of course. And even if it's not love at first sight, you can help turn the first date with a customer or supplier into a lifelong romance.

Free Riders In 1981 the US airline industry was experiencing turbulence. The market was still adjusting to deregulation. A wave of entrants – mainly no-frills carriers, such as People Express – was adding new capacity. The established airlines found themselves in a fierce battle for passengers. Prices tumbled. The airlines discovered that passengers, in pursuit of low prices, displayed no loyalty.

At least that was the game until May 1 of that year, when American Airlines unveiled its AAdvantage frequent-flyer program.[25] The program allowed passengers to build up credits for each mile flown and then redeem these credits for free flights to Hawaii and elsewhere. The more miles traveled, the better the reward. Each flight on American created a greater incentive to stick with American for the next one. American's customers thus became loyal.

With the help of plastic membership cards, a computer program to track miles, and some empty seats, American had created loyalty out of thin air. Even better, AAdvantage was most valuable to American's most profitable customers – frequent business travelers.

AAdvantage created loyalty by rewarding it – a fine idea, as long as it's not too costly. The frequent-flyer program was *very* cost-effective. What was the cost to American of providing the free-flight award? Some peanuts and a little fuel – about $20. Just as TWA's Comfort Class took out empty seats to give passengers more legroom, American used otherwise empty seats to give loyal customers free travel.

Of course, there was some cannibalization. Some of the people who got free trips to Hawaii would have paid to go there anyway.

In these cases, the cost to American was more like $1,000 than $20. The traveler saved $1,000 he would have paid, and American lost $1,000 it would have collected. Even here, AAdvantage bought loyalty – in fact, $1,000 of loyalty – but the cost was a dollar per dollar of loyalty created.

But most of the time there wasn't cannibalization. Awards went to people who were very happy to fly to Hawaii for free but wouldn't have gone if they'd had to pay $1,000. For argument's sake, let's say the person valued the trip at $500. Then American bought $500 of loyalty at a cost of $20. Half as much loyalty at one-fiftieth the cost – much more cost-effective.

There were two caveats, however. First, American had to ensure that free travelers didn't displace other, fare-paying passengers. If the free trip bumped a fare-paying passenger because the plane was full, then the cost to American was again $1,000. So American restricted the free travel by limiting the number of seats on each plane available to award holders and imposing blackout dates. Second, American had to prevent resale of the awards. Otherwise, they would have likely ended up in the hands of people who would have paid full fare. And once again, the cost to American would have been $1,000.

Of course, customers would have found the frequent-flyer program more desirable without these restrictions. So why didn't American listen to the customer? Because then the cost would have been closer to $1,000 than $20, and the program might not have been worth doing. The restrictions weren't there to annoy people; they were essential to make the program fly.

Some economists have asserted that a frequent-flyer program doesn't add value. In an op-ed piece in the *Los Angeles Times*, Oxford University professor Paul Klemperer and UCLA professor Ivan Png argued: 'The customer loyalty built by frequent-flyer programs is . . . not based on enhanced satisfaction of consumer wants, unlike the development of a milder soap or a more fuel-efficient sports car.'[26]

We disagree. American spotted some underutilized resources – empty seats – and put them to good use. People got to take vacations in Hawaii that they otherwise wouldn't have. In this way, American made the pie bigger and raised its added value.

AAdvantage American. But there was nothing in the program that other airlines couldn't copy.

Divided Loyalty Only two weeks after American introduced AAdvantage, United Airlines launched its own Mileage Plus frequent-flyer program. Within three months, all the major US carriers had frequent-flyer programs of their own. Even so, AAdvantage continued to lead the way with more than 1 million members enrolled by the end of 1981.

Not everyone could duplicate the AAdvantage program equally well. American was way ahead in using computer technology, giving it a significant edge in running its program. Airlines without the capability to automate found the programs much more cumbersome to administer. To track miles, they had to collect coupons presented to gate agents prior to boarding.

Frequent-flyer programs gave the large airlines an advantage over the regional carriers and start-ups. With more extensive route systems, they could offer free flights to Hawaii, the Caribbean, and other tempting destinations. Regional airlines tried to compensate by offering more generous rewards. In turn, the major carriers further enhanced the appeal of their programs by forming alliances with international carriers.

How bad was it for American that its AAdvantage program was copied so quickly? It's true that imitation reduced American's ability to take share. American was no longer unique. All the airlines now offered improved products, so the playing field was again more or less level.

But that doesn't mean that the loyalty effect was lost. Even with all the other frequent-flyer programs out there, after a customer racks up a few miles on American, he has an incentive to stick with American the next time. That's the real genius of frequent-flyer programs. Even when copied, they still create an incentive for loyalty.

Just as American has loyal passengers, now all the other airlines do, too. And once each airline has its own base of loyal travelers, going after share through low prices becomes less attractive. Suppose United lowers price in an attempt to take market share. Doing so is less effective because it's harder to attract American's frequent flyers. Similarly, if United raises price, it won't lose as many of its passengers, because they don't want to abandon their miles on United.

Overall, price cuts are less effective, and price rises less risky. This is true for every airline. Lowering price gains fewer new customers; raising price loses fewer existing ones. The loyalty effect is particularly strong in the business travel market, where passengers have the most miles. That's why unrestricted-coach and business-class fares have been so stable.

Frequent-flyer programs do have their flaws. People can join more than one program, and this dilutes the loyalty effect. Even so, there is an incentive to concentrate miles on a few airlines. One 40,000-mile award is better than two 20,000-mile awards, so it doesn't pay to spread miles across too many airlines. Another drawback is that when miles are cashed in, the loyalty effect fades.

Both of these flaws have been corrected with the airlines' introduction of Gold and Platinum programs. These second-generation programs give VIP privileges to the airlines' best customers. To attain Gold status with American requires flying 25,000 miles a year, Platinum status comes with 50,000 miles. United even has a special category − 1-K Fliers − for those who rack up over 100,000 miles a year on United. Because the cutoff levels are so high, it's rare to be able to be a member of more than one Gold or Platinum card program. Thus, the loyalty effect doesn't get diluted.

Once people hit the cutoff levels, they enjoy unlimited first-class upgrades, special customer reservation service, companion upgrades, and more, over the next twelve months. It's not a onetime reward, so the loyalty effect doesn't fade. To take advantage of these privileges and maintain their Gold or Platinum card status, travelers have every reason to continue flying the same airline. The Gold and Platinum card programs are important because they appeal to the airlines' most precious customers − those who pay full fare and fly nonstop.

Who Wins and Who Loses? Frequent-flyer programs create loyal customers, and that leads to a win-win for the airlines. They're an example of what we meant when we said that sometimes the best way to succeed is to let others do well, including your competitors.

What about customers? They get a free trip to Hawaii but pay extra for all those business trips to Philadelphia. Thus, there are two effects going in opposite directions. On balance, most business travelers consider themselves better off because the boss pays for the ticket and they get the miles. So are bosses the losers? Not necessarily. Frequent-flyer miles are a tax-free way for companies to compensate employees who undertake a lot of business travel.

Actually, the win-win outcome for the airlines is a controversial point among some. For example, Professor Max Bazerman of Northwestern's Kellogg School of Management argues that frequent-flyer programs are an instance of a mutually destructive escalation game. He points to episodes of intense competition, such as the 1987 'triple miles' war, when American, then Delta, and then everyone else offered three frequent-flyer miles for each mile traveled.[27] Bazerman observes that, as a result, travelers have stockpiled well over a trillion frequent-flyer miles – a potentially huge hidden liability for the airlines. His conclusion: frequent-flyer programs have been a lose-lose game for the airlines.

We agree that a trillion (1,000,000,000,000!) frequent-flyer miles is a very big number. A back-of-the-envelope calculation suggests that clearing the slate would require nearly 100,000 roundtrips of a full 747.[28] That's a lot more than some free peanuts.

But that perspective greatly overstates the real cost. To avoid displacing paying customers, airlines restrict usage. As a result, many miles will simply never be used. Furthermore, airlines, somewhat controversially, changed some of the program terms: the miles can expire if unused within a certain time; the number of miles needed for free travel has risen. The reality is that the airlines have a lot of discretion in cutting the liability down to a manageable size.

It's true that frequent-flyer programs create competition among airlines to sign up new members. Hence the triple-miles war. It's worth investing today to create loyalty tomorrow, but only to a point. It isn't smart for one airline to fight too aggressively to sign up members. It should prefer that other airlines have some loyal customers, too. Otherwise, competitors have little to lose; they have every reason to cut price in order to attract customers. Remember the glass-house effect described in the previous chapter under 'Eight

Hidden Costs of Bidding.' It's important that competitors have something to lose from getting into a price war.

Even with their flaws, frequent-flyer programs are a stroke of genius. They are an unparalleled way of increasing the total pie and creating loyal customers. True, competition to sign up members may, at times, have escalated out of control. But, on balance, we think it's been a win-win-win – for American, for the other airlines, and for passengers, not to mention tourism in Hawaii.

Frequent-flyer programs have helped companies in other businesses engineer customer loyalty, too. Car rental agencies, credit-card issuers, hotels, long-distance phone companies, and others have formed 'affinity' programs with the airlines to allow them to offer frequent-flyer miles to their own customers. For example, there's the Citibank credit card with American, and the First Chicago card with United. Each dollar charged on these cards earns a frequent-flyer mile. Long-distance phone company MCI gives five miles on Northwest for each dollar of calls; sign up with Sprint and get five miles on TWA. Program partners pay the airlines a fee for this privilege, around a penny a mile. These pennies add up. The airlines earn an estimated $2 billion annually in fees from affinity programs. The programs are so large that over half of the miles credited to frequent-flyer programs are now earned on the ground.

Saying thank you

Frequent-flyer programs hold an important lesson for every business. You want to say thank you to your loyal customers. It's a critical step in building a relationship. We think that saying thank you is an essential part of establishing added value in a competitive market.

Say thank you.
Create loyalty by rewarding it.

Affinity programs are one way to say thank you to your customers. But you may want to set up your own loyalty program. Hotel chains offer their frequent guests guaranteed reservations, upgrades to executive-tower suites, and more perks. Car rental agencies, rail-roads, casinos, and others all have their own programs. We think every business should have a loyalty program.

Of course, there are better and worse ways of saying thank you to your customers. Our experience working with companies to develop loyalty programs has helped us come up with some guide-lines on how to say thank you most effectively.

1. Say thank you in kind, not cash. The most cost-effective way to say thank you is to reward customers in kind, not cash. Imagine if the airlines promised travelers a $500 savings bond as a reward for 40,000 miles of flying.[29] Customers would value this at precisely the amount it cost the airline to buy. It's a large thank-you, but it's also an expensive one – each dollar of thanks would cost the airline a dollar. A much more cost-effective way to say thank you is to give travelers something that they value at $500 and that costs much less than that to provide. The airlines accomplish this by using empty seats to give away free trips.

When you use your own product to say thank you, it's cheap to provide in comparison with its value to the customer (or else you shouldn't be in business in the first place). This is especially true if you have excess capacity. Even if you don't actually give away your product for free, giving it away at cost or at a discount is still effectively saying thank you in kind, not cash.[30] Done right, saying thank you doesn't have to cost you much.

People understand the kind-not-cash principle very well in some contexts, but then forget about it in others. For example, insurance companies sometimes replace losses with product rather than cash. It's more cost-effective this way. With their experience and clout, insurance companies can often replace damaged or lost goods at a lower cost than the customer could. To us, that says that insurance companies should also be helping their loyal customers buy new products. It would be a cost-effective way of rewarding loyalty. For example, auto insurers should help their customers buy new cars, a strategy we came at from a different direction in the previous chapter.

The same idea applies in business-to-business relationships. You may be able to say thank you to a customer by helping it get a better price on its raw materials. You use your superior clout or access to help your customer. Whether your customers are consumers or other businesses, look for ways to say thank you in kind.

2. Save the best thank-you for your best customers. Many companies offer the best deals to new customers. That's backward. You want to treat your best customers the best.

New customers are an unknown quantity. You don't know how profitable they'll turn out to be. With your existing customers you know how things turned out. Figure out who are your most valuable customers and don't ever lose them. Reward them to ensure their loyalty. It's like remembering to send flowers even after you've turned a romance into a relationship.

The cellular-phone carriers could take this message to heart. There is currently tremendous excess capacity in the business that could be used to create much more loyalty among customers. For example, carriers could give their best customers free weekend and evening calls. Or they could give customers free incoming calls in proportion to the time spent making outgoing calls.

Some cellular carriers do give free weekend and evening benefits, but only to new customers. Others offer fantastic discounts on the new digital phones, but again, only to new customers. Wrong approach. This practically invites customers to keep switching service back and forth to get the promotions. It actually undermines, rather than builds, loyalty.

Once again, frequent-flyer – and especially Gold and Platinum card – programs get this right. The best customers are treated the best.

3. Say thank you in a way that builds your business. Often, businesses give potential customers a free trial. That's a fine strategy for increasing sales. But even better may be to give loyal customers a guest pass and encourage them to bring someone else along. Health clubs often do this.

You can also reward loyal customers with products that encourage them to do more business with you themselves. For example, the long-distance phone companies could reward loyal customers by giving them free, or at least sharply discounted, voice mail. With voice mail, no call goes unanswered. Thus, the phone company doesn't lose the business because someone wasn't home. Even better, someone has to make a call to pick up the message and perhaps even another call to answer it – or leave voice mail in return. Just as voice mail solves the problem of no answer, call waiting solves the problem of a busy signal. Now more calls get through, and so again there's more business. We suspect that giving free three-way calling to loyal customers would further stimulate phone usage.[31]

4. Don't say thank you too quickly, or too slowly. If you say thank you before you've had enough time to build a relationship, it's more like offering a discount up front. Similarly, if you wait too long to say thank you, it won't mean much.

5. Say that you're going to say thank you. Yet another well-designed feature of frequent-flyer programs is that they tell you right from the start how big a thank-you to expect. The more traveling you do, the bigger the thank-you you'll get. With some other programs, the thank-you is a surprise – a welcome one, but still a surprise. For example, American Express recently sent one of us a reward for being a ten-year member. Nice, but unexpected. And we don't know whether there'll be another thank-you down the road for being a fifteen- or twenty-year member.

If you're planning to say thank you, you should let your customers know in advance so they'll stick around for it. You don't have to treat your customers as though you're throwing them a surprise birthday party. You might be the surprised one if they don't show up. Telling them that you have a present waiting for them down the road doesn't ruin the sentiment.

6. Recognize that you may have to compete for loyalty. Saying thank you to customers is a way to make them loyal to you. Of course, you may have to compete with others who'd like those same customers to be loyal to them.

One way to compete is to give deep discounts to attract new customers. The problem with this strategy is that others will be forced to do the same, and then some of your existing customers will leave to become someone else's new customers. The result is likely to be no net shift in share, simply an expensive reshuffling of customers and a loss of loyalty all around.

A better way of competing for customers is through creating a more attractive loyalty program. That way, the end result of competition for customers is more, not less, loyalty. But don't go overboard on this, either. That's our next point.

7. Allow your competitors to have loyal customers, too. Just as you benefit from having loyal customers, don't begrudge others this opportunity. This isn't just a matter of being nice. Consider the alternative. If your competitors don't have loyal customers, they certainly won't have an incentive to raise price. Indeed, with little to lose, they'll be more likely to cut price in an attempt to attract your loyal customers. It's to your benefit if your competitors develop their own loyal customer bases. As we've said, rivals who live in glass houses don't throw stones.

8. Don't forget to say thank you even if you have a monopoly. You may have a monopoly today, but there's always the possibility of entry or competition from a new technology tomorrow. Having loyal customers will put you in a better position in tomorrow's game. It may even dissuade others from entering your game altogether. The best time to develop loyalty is while you have a monopoly. You don't have to compete for customers; they're all yours.

Cable TV companies really seem to have missed the boat on this. They have the ability to give their loyal customers premium services that are currently sitting idle. Very few people subscribe to all three of HBO, Showtime, and Cinemax. People would like to have a second or third premium channel, but with the overlap among them, it's often not worth the high price. But it's worth something. So after someone has subscribed to HBO for a year, the cable company should throw in Cinemax for an extra dollar. Or after people have paid for ten pay-per-view movies, the cable company should give them a selection of movies they can view for free.

Why have cable companies missed the boat? Perhaps because they have the monopoly mindset. They don't feel the need to create loyalty. Big mistake. Their markets are opening up. Today they face nascent competition from satellite TV and 'wireless' cable. Tomorrow they'll wish that they'd invested more today in creating loyal customers. The local phone companies, like the cable companies, are also facing deregulation of their industry. And they're making the same mistake: they're failing to create customer loyalty while they have the game all to themselves.

9. Say thank you to your suppliers as well as to your customers. Recall the symmetry of the Value Net: every strategy toward customers has a symmetric counterpart toward suppliers. Just as you should reward loyal customers, you should reward loyal suppliers.

Is this really necessary? You are your suppliers' customers. Isn't it incumbent on them to have loyalty programs for you? Why do you need loyalty programs for them? The answer is that the programs should go both ways. You want your suppliers to be loyal to you – that's why you need a loyalty program. Your suppliers want you to be loyal to them – that's why they need loyalty programs.

Companies understand the idea of saying thank you to suppliers as it applies to employees. They often give employees freebies, or at least large discounts off the company's products. Many companies offer recognition, even rewards, to their longtime employees. But companies should think beyond their employees and reward other loyal suppliers, too. If you offer employee discounts on your products, think about extending this benefit to all your suppliers (and their employees). Or you may be able to use your superior clout and access to help a supplier get a better price on its raw materials. These are just two ways of saying thank you to your suppliers.

Every company should have a frequent-customer program and a frequent-supplier program. This is the way to create loyalty both up and down the Value Net.

NINE TIPS ON SAYING THANK YOU

1. Say thank you in kind, not cash.
2. Save the best thank-you for your best customers.
3. Say thank you in a way that builds your business.
4. Don't say thank you too quickly, or too slowly.
5. Say you're going to say thank you.
6. Recognize that you may have to compete for loyalty.
7. Allow your competitors to have loyal customers, too.
8. Don't forget to say thank you even if you have a monopoly.
9. Say thank you to your suppliers as well as to your customers.

---

4 Imitation

The American AAdvantage program was our model example of how to create customer loyalty. We saw how AAdvantage worked and how it continued to work even after the other airlines copied it. Today practically every airline has a frequent-flyer program. Yet the programs remain an effective strategy for creating loyalty. A strategy that continues to work even after being copied? This seems to fly in the face of conventional wisdom.

Conventional wisdom has it that imitation is the bugaboo of business strategy. You develop a new strategy. It works. In fact, it works so well that everyone copies you. And then your strategy no longer works. You have to come up with something new again.

That's the view commonly expressed in business-strategy textbooks. Yale School of Management professor Sharon Oster writes: 'If everyone can do it, you can't make money at it.'[32] Bruce Henderson, founder of the Boston Consulting Group, compares the effects of imitation in business to a biological phenomenon called Gause's Principle of Competitive Exclusion: 'No two species can coexist that make their living in the identical way . . . no more in business than in nature.'[33]

It's true that when someone literally copies your product or processes, that erodes your added value. Thus, American, with its 10:00 P.M. departure from Los Angeles to New York, loses some added value when United offers a 9:55 P.M. departure. Now American isn't something so special in the air.

Some people take the ill effects of imitation even further. They say that there can be no long-lasting formula for generating successful business strategies.[34] Why not? Because if someone discovered such a formula, the formula would be most unlikely to remain secret for long. And then any strategy generated by the formula could, and would, be copied. No one could make more than temporary gains. If correct, this argument would imply that there's a limited shelf life for any business prescription.

But this argument isn't correct. Take frequent-flyer programs: everyone can do it, everyone can make money at it, and everyone can continue making money at it. The gains are more than temporary. This isn't just true for frequent-flyer programs. There are other strategies, such as meet-the-competition clauses and rebate programs, that also continue to work after being copied. (We'll examine these strategies in the Rules chapter.) So where's the mistake in the conventional wisdom on imitation?

Healthy imitation

Imitation is the bugaboo of business strategy if the goal is to secure 'competitive advantage' – to do better than others. Imitation means everyone can do the same thing. And then you can't do better than others, since they are doing just what you're doing. It's almost tautological that you can't have a sustainable competitive advantage.

Imitation is harmful whenever you think win-lose. You discover a win-lose strategy. You take two steps forward, and your competitor takes one step back. If your competitor can imitate you, the strategy gets turned around. Now it's lose-win. Your competitor takes two steps forward, and you take one step back. Here, imitation is harmful: it erodes your initial gains.

That's if you are lucky. It can get much worse than that. Sometimes win-lose means one step forward for you and two steps back for your competitor. The classic example of this is competing on price.

You go after a competitor's customers through low prices. When you do that, you gain less than your competitor loses because you gain a customer at the new, lower price, while your competitor loses a customer who was paying the old, higher price. It's one step forward for you and two steps back for your competitor. You're ahead, but only until your competitor responds by going after one of your customers. Now it's one step forward for your competitor and two steps back for you. The net result is one step back for both you and your competitor. Prices are lower, and market shares end up back to where they were. Now imitation isn't just harmful, it's deadly:

win-lose + lose-win → lose-lose

So yes, win-lose strategies can backfire – quite badly – because of imitation.

Where conventional wisdom goes wrong is in ignoring the possibility of win-win strategies. Not a surprise, given the conventional mindset of business-as-war. With win-win, imitation is healthy. Now it's one step forward for you and one step forward for your competitor. After imitation, it's another step forward for your competitor and another step forward for you. Imitation actually amplifies the gains:

win-win + win-win → WIN-WIN

Look at frequent-flyer programs again. American got an advantage from AAdvantage. The program gave American a way to attract some passengers from other airlines. So far win-lose. One step forward for American and one step back for United. When AAdvantage was imitated, American lost its ability to gain share. One step forward for United and one step back for American. A wash.

But there was also a win-win element when AAdvantage was introduced. With more loyal customers, American wasn't about to start a price war. Indeed, it could even raise price. That gave the other airlines some room to raise price, too. Along with the win-lose component in terms of share shifts, there was a win-win element in terms of pricing.

Imitation of AAdvantage eliminated the win-lose component and reinforced the win-win effect. We saw that when every airline has a frequent-flyer program, customers are more loyal. Price cuts are less effective and price rises less risky. There's less incentive to compete on price. The result is greater price stability, especially in the business travel market.

In sum, imitation of win-win strategy is healthy, not harmful. So if you come up with a win-win strategy, you don't have to keep it secret. It's not a problem if your strategy becomes widely known and widely imitated. In fact, that's all to the good. The more competitors that adopt your strategy, the better for you.

By the same token, in writing down the theory of loyalty programs in this book, we haven't undermined their effectiveness. On the contrary, we hope that what we've written will prompt more companies to set up loyalty programs – and the more that do, the better.

Unhealthy imitation

Not everything is win-win. You have to be prepared for the possibility of unhealthy imitation. What can you do if gains are temporary? You need to create a long-lasting series of temporary gains.

The trick is to run faster and faster. You make a better product. Others then copy you. But by then you're a step ahead. You've already improved your product.

The game isn't about how good your products are; it's about how good you are at improving them. It isn't where you are; it's how fast you're moving. It isn't position; it's speed. You never stand still; you're a moving target.[35]

What if others copy your improvement process? They become as good as you at improving products. What then? You've already improved your improvement process. Now the game isn't about how good your products are, or even how good you are at improving them. It's about how good you are at improving your improvement process. It's not about where you are or even how fast you're moving. It's how fast you can speed up. It's not about position or speed. It's about acceleration.[36]

And, in principle, there's even improving how you improve your improvement process, and so on. Pretty rarefied stuff? Not to Individual, Inc.

What Makes Individual, Inc. Unique Individual, Inc., provides a hightech clipping service to help people deal with the information explosion. Clients tell Individual, Inc. what topics they're interested in. The company then uses computer searches to find all the relevant articles, ranks them, and makes them available via a personalized web page, email, or even old-fashioned fax. Clients get full text, abstracts, or headlines of articles, according to their priority ranking. The result is like getting your own private news briefing.

Individual's problem is how to preserve its added value. What prevents other companies from copying this service? Although Individual's SMART (System for Manipulation and Retrieval of Text) software is proprietary, its basic concept is not. But Individual, Inc. has gone the next step. It regularly asks clients to evaluate each story they're sent on a 'not relevant' to 'somewhat relevant' to 'very relevant' scale. This 'relevance feedback' is fed into SMART, which uses it to improve the selection of stories delivered to each client. SMART also learns from watching what clients read, not just listening to what they say. Every time a client retrieves a full-text article from an abstract or headline, SMART records this fact and updates its user profile accordingly.

A typical client might start by evaluating 50 percent of the stories as very relevant. Individual, Inc. can get the score up to 90 percent within a month. Not only do clients get an improved product but they also see a very tangible measure of the improvement. That's a useful reminder to Individual's clients that they'd go back to 50 percent if they were to switch to another clipping service.

Individual, Inc. has been improving not only its product but also how fast it can improve its product. It has developed a better ability to interpret feedback. Experience has allowed it to fine-tune the parameters of its learning algorithm, and so it learns about a client faster now than when it began the service.

By learning about its clients' preferences, and by learning how to learn faster, Individual, Inc. preserves its added value.

The Individual, Inc. strategy can also work for bookstores, magazines, video rental stores, even dating services – indeed, for any business where the trick is matching a wide range of options to diverse customer wants. Thus, when a bookstore gives a customer a recommendation, that's an opportunity to follow up and discover why the book did or didn't please. That way, the bookstore can make a better recommendation the next time. A rival store can sell the same book, but without the feedback the first store has, it can't replicate the good advice. At least, it can't until it has also learned about the customer. And if the customer is happy with the first store, it may not get that chance.

The *Harvard Business Review* asks its readers to rate each article. The magazine uses this feedback to improve its selection of material. The result is a magazine better tailored to its readers, who, in turn, become more loyal. We want to develop a loyal readership, too. That's why we hope you'll give us feedback on this book. (Our Internet addresses are on the first page of this book.)

Rapid product improvement works for Individual, Inc. It can work for bookstores and magazines. It certainly works for Intel, which has a strategy of jumping to the next-generation chip even before its competitors have finished cloning the current generation. But it isn't a panacea. With some products, it simply isn't realistic to think that you can keep coming up with improvements.

Take soap. Here, you're lucky if you come up with one innovation. There aren't that many ways of improving soap. This was a very real problem for Minnetonka, a small Minnesota company that came up with a new type of soap.[37] How could Minnetonka prevent the likes of Procter & Gamble and Lever Brothers from copying its idea?

Pumping Up Profits In 1964 entrepreneur Robert Taylor took $3,000 of his savings and founded Minnetonka. Over the following two decades, Taylor took the company from being a niche producer of novelty toiletries to being a market innovator and player in the soap, toothpaste, and personal fragrance businesses. Among Minnetonka's widely known products are Softsoap liquid soap, Check-Up toothpaste, and the Calvin Klein fragrances Obsession and Eternity.

Taylor wasn't afraid to experiment. One spring alone, Minnetonka brought out seventy-eight new products. It's the 'spaghetti test' strategy. You throw things up against the wall and see what sticks. Tom Peters calls this the 'antistrategy' strategy: don't analyze, just see what works and do more of it. Taylor was essentially test-marketing a whole series of products, looking for a 'hit' product with which to enter the mass consumer market. And he found one.

In 1977 came the Incredible Soap Machine, which dispensed liquid soap from a plastic pump bottle. It quickly became Minnetonka's best-selling product. Minnetonka was onto something. The Incredible Soap Machine eliminated the puddle of soap ooze messing up guest bathrooms across America. Here was an opportunity to enter the mass market for soap.

The Incredible Soap Machine was renamed Softsoap, and sales began to take off. In test markets, Softsoap captured between 5 and 9 percent share of total bar-soap sales. Softsoap was ready to go national, and in 1980 Minnetonka launched a $7-million advertising campaign. To put this in perspective, Armour-Dial's advertising budget for Dial soap, the market leader, was $8.5 million.

Softsoap repeated its test-market success, reaching $39 million in sales the first year. Taylor believed that the liquid-soap market would reach $400 million. But what share could he keep once the leading bar-soap manufacturers came out with their own versions of Softsoap?

The threat of imitation was real. Softsoap was hardly a patentable innovation. Pumps have been around since Archimedes. The brand name was good, but there were other, more established brand names in the soap business. The Ivory name, for one, was over a hundred years old.

Taylor had experienced the problem of imitation before. He had tried bringing fruit shampoos to the mass market: 'I can remember when I saw the Clairol knock-off of our shampoo . . . We create the concept and they take it to the marketplace . . . When they moved theirs into drugstores and food stores, our line just withered.'[38] How could he preserve any added value once the likes of Armour-Dial, Procter & Gamble, Lever Brothers, and Colgate-Palmolive muscled in with their brands and distribution clout?

Taylor had a window of opportunity. The majors adopted a wait-

and-see stance. Liquid soap was still unproven and they preferred to let Softsoap be the guinea pig. Once they saw that Softsoap was a success, the majors decided to do their own test-marketing. A surprise awaited them.

An essential part of any liquid-soap product is the little plastic pump, and Taylor realized that there were just two suppliers. In a bet-the-company move, he locked up both suppliers' total annual production by ordering 100 million pumps. Even at 12 cents apiece, this was a $12-million order – more than Minnetonka's net worth. Capturing the supply of pumps gave Taylor another eighteen to twenty-four months. Softsoap had bought some more time to build brand loyalty and thus establish its added value.

When Procter & Gamble finally test-marketed a liquid soap, it used the name Rejoice rather than risk sullying the Ivory brand. The results weren't encouraging, so Procter & Gamble again delayed. Through a combination of pluck and luck, Taylor had a three-year window. It wasn't until 1983 that Procter & Gamble finally introduced its Ivory brand liquid soap, which quickly gained almost 30 percent of the liquid-soap market. Jergens liquid soap from American Brands was a distant third. By 1985 the liquid-soap market had grown to $100 million. Softsoap maintained its lead position with a very respectable 36 percent share. Two years later, Colgate-Palmolive, which had missed the boat on liquid soap, played catch-up by buying Softsoap for $61 million.

There wasn't anything Minnetonka could patent. But even a patent doesn't completely protect you from unhealthy imitation. You have to plan for the day when the patent expires or someone comes up with an alternative solution. In the NutraSweet-Holland Sweetener story back in the Players chapter, we saw how NutraSweet used the period of patent protection to prepare for future imitation. It made big investments in branding its product and in moving down the learning curve. Once the patent expired, Holland Sweetener couldn't match NutraSweet's product or cost position. NutraSweet also made it harder for Holland to catch up. By aggressively fighting Holland's entry in Europe, NutraSweet denied Holland volume and thereby slowed its progress down the learning curve.

ANTIDOTES TO UNHEALTHY IMITATION

1. Collect customer feedback to customize your product – competitors can't copy you because they don't have the information.
2. Create a brand identity.
3. Build volume to move down the learning curve.
4. Compete aggressively for volume so that competitors can't follow you down the learning curve.

Individual, Inc., Minnetonka, and NutraSweet faced clear threats from unhealthy imitation, and adopted effective strategies to protect their added values. Sometimes, though, the threat of unhealthy imitation is less apparent, and then companies may inadvertently adopt strategies that make them more vulnerable. We give two examples. In the first, we were asked to disguise the identities of the players, so the names are fictional.

Outboxed Andy had a breakthrough day. For years, his plant had been supplying Polymatic, a key customer, with specialty cartons for its consumer products. But it was a crazy way of doing business. Polymatic had four divisions: each had its own different set of specifications, and each handled its own purchasing. As a result, Andy had to develop separate products and respond to many separate orders, which led to high set-up costs and short runs.

Andy realized he could save some money by doing things differently. He could do a better job adapting to Polymatic's quirks. Using longer runs and holding more inventory himself might save perhaps 2 cents a square foot. That was the small news. The big news was that even greater savings could be realized if Polymatic also agreed to do things a little differently.

There was really no need for all the variety. If some of the different divisions at Polymatic would go along with the specifications used by the others, there would essentially be no loss at all in quality. As things stood, each division ordered material as and when its inventories fell to a certain level. It was all done by a computer program designed to minimize Polymatic's costs, not Andy's. The program design didn't take into account the possibility that if

Polymatic placed larger orders and held more inventory, Andy could offer a lower price.

If Polymatic would standardize its specifications, coordinate its ordering, and hold more inventory itself, Andy could do significantly longer runs on fewer cartons. He judged the combined savings, net of Polymatic's extra inventory cost, to be 10 cents a square foot. That would add up to several hundred thousand dollars.

Of course, that led to a bit of a dilemma. He could save 2 cents on his own in ways that would be invisible to Polymatic. But to create the 10-cent savings, he'd need to get the different groups at Polymatic to standardize, coordinate, and change inventory policy. That would require sharing his analysis with Polymatic, and some of the cost savings, too.

Andy decided to go for the big savings. He went to Polymatic, explained the new approach, and offered to split the savings. He figured that was ample incentive for Polymatic to change its way of doing business. Polymatic would appreciate the windfall, which in turn would lead to friendlier negotiation of next year's contract. It would create goodwill and prove his 'partnership' orientation.

Polymatic appreciated Andy's initiative. Six months remained on the existing contract, and Polymatic was happy to save a nickel. Everything looked like a win-win.

The first indication of trouble came at the contract renewal. Polymatic sent out a request for bids and got four new bidders. In the past, there had never been more than one other bidder. Andy's company was uniquely qualified to provide all of the specialty cartons and small production runs at a reasonable cost. Although this niche production was inherently inefficient, Andy was a very efficient niche producer.

What had changed? According to Polymatic's purchasing agent, now that the specs were standardized and production runs were longer, the big players in the container market were, for the first time, showing interest in the account. Andy was floored when Polymatic came back and said that one of the larger producers had come in with a bid that was 20 cents a square foot below last year's price. The purchasing agent said that he appreciated Andy's help in making the cost savings possible, and if Andy was willing to meet the new price, he could keep the business.

Andy had no alternative. He took the hit. The profit margins were large enough, just, to make it worthwhile. Still, Andy wondered how he could come up with a way to cut costs by 10 cents a square foot and end up giving a 20-cents-a-square-foot price cut.

Andy had inadvertently lowered his own added value. It's true that he cut the cost of supplying Polymatic, but the new approach was completely imitable. For companies that were not set up for niche production, Andy had managed to cut the costs of serving Polymatic by so much that they now found it profitable to enter the game. The result was a terrific win for Polymatic — and not so great for Andy.

What should Andy have done? Perhaps he could have used a long-term contract. But, ultimately, this was not going to be a long-run win for him. The real issue is that Andy had added value as a niche producer. But lowering the cost of supplying Polymatic made what he did much easier to imitate. Whether or not it was in the customer's interest, it wasn't in Andy's interest to change the way the customer ran its business. It may have happened of its own accord sometime in the future, but there was no sense hurrying it along.

Andy figured out what the customer wanted, shared the knowledge, and then watched the customer take the knowledge to the competition. He made a larger pie that others could eat. That's a mistake that even the most famous companies have made. We believe it's exactly the mistake IBM made when it entered the personal computer business.

Many people have written about the difficulties IBM ran into in the late 1980s and early 1990s. In particular, they've commented about IBM's failure to translate the towering position it once had in mainframe computers into a strong position in personal computers. Some people point to concerns at IBM that personal computers would cannibalize the mainframe business. Others point to IBM as a textbook example of the difficulties large organizations have in making big changes in the way they do business. Let's look at the IBM story from the perspective of unhealthy imitation.

Loss of Computing Power When IBM entered the personal computer market in 1981, Apple Computer was ahead and IBM was playing catch-up. Speed was all-important. At that time, IBM needed to establish an installed base of its machines, and fast. It wanted to move from design to market in just twelve months. To this end, it abandoned its tradition of internal development and, instead, turned to Intel and Microsoft to supply the microprocessor and operating system technology for its PC. IBM was lauded for this innovative move.[39]

The benefit of IBM's outsourcing decision was the rapid creation and adoption of the IBM PC. There was a larger pie sooner. The cost of its decision was having to share the pie with Intel and Microsoft. For the sake of simplicity, let's assume that IBM, Intel, and Microsoft could each get one-third of that larger pie. For IBM, that might well have been better than a much larger slice of the much smaller pie that would have resulted from keeping everything in-house. For Intel and Microsoft, of course, sharing the big pie was just fine.

Outsourcing was half of IBM's strategy in PCs. The other half was an open-architecture policy. The rationale was to help programmers write application software for the IBM PC. An unintended consequence was that other players started cloning IBM's hardware. They quickly resolved any software incompatibility problems. First Osborne, Leading Edge, and Hewlett-Packard, then Compaq, Dell, and hundreds of others entered the hardware business, all building IBM clones using Intel chips and Microsoft operating software.

Now Intel and Microsoft were more important than ever. Microsoft was a monopoly supplier of an essential input into the fast-growing business of making IBM-compatible PCs. Although Intel was forced to license its original 8086 chip, with each successive generation Intel became more of a monopoly supplier, too. The bigger the pie, the better for Intel and Microsoft. But the implications for IBM were very different. A Compaq machine was a very decent alternative to an IBM. No longer the sole provider of hardware, IBM now had a drastically curtailed added value.

IBM's real error was pursuing the outsourcing and open-architecture policies together. Had it stopped at bringing in Intel and Microsoft, and not given up control of the hardware portion of the business, it would have remained in a strong position. Had it kept control over the chip and operating system technologies, then, despite cloning of the hardware, it would still have been in a strong position. Either approach might well have been effective. But out-sourcing together with opening the architecture was a mistake. It's a case of two rights making a wrong.[40]

Even bringing in Intel and Microsoft and opening up the architecture wouldn't have been as bad for IBM had it made Intel and Microsoft pay to play. IBM could have demanded equity stakes in Microsoft and Intel in return for bringing them in. Back in the early 1980s IBM had the power to shape the game in this way. But it missed these opportunities.

Realizing its error, in 1987 IBM tried to regain control by introducing the PS/2 line of personal computers, which contained the OS/2 operating system developed jointly with Microsoft. But by then it was too late. Microsoft didn't need IBM. Microsoft's 1990 release of Windows severely damaged the prospects of OS/2 becoming widely adopted.

If you do give up much of the pie to another company, the smart strategy is to acquire a piece of that company. Early on, IBM had the financial resources to take a large equity position in both Intel and Microsoft. It did make an investment in Intel in 1982, getting a 20 percent equity stake and warrants to purchase another 10 percent. But, ironically, IBM sold out in 1986 and 1987 for $625 million. Ten years later that 20 percent stake would have been worth $25 billion. As for Microsoft, IBM had an opportunity to buy 30 percent of Microsoft for under $300 million in mid-1986. By the end of 1996, that $300-million investment would have been worth $33 billion.[41] Had IBM acquired and kept large equity positions in both Intel and Microsoft, we suspect that today, people would be talking about IBM's continuing success rather than its failures.

Our first and last stories in this chapter are polar opposites. Nintendo exerted tight control on the video game business. Had Nintendo adopted an open-architecture policy and not put in a security chip, perhaps the total pie would have been larger. Perhaps not. There might simply have been a flood of low-quality games into the market and another Atari-like crash. Either way, Nintendo's actions ensured that its added value was equal to the whole of the pie, whatever its size, and that everyone else's added value was kept small. In contrast, IBM's outsourcing and open-architecture policies cost it control of the PC business. By facilitating more rapid and widespread adoption of the IBM PC platform, IBM increased the size of the pie. But by erasing much of its own added value, IBM sharply limited its ability to capture the pie.

Once again, the distinction between the added value of a company and a product is crucial. Just as the added value of Sony is much less than that of televisions, and the added value of Nissan much less than that of cars, so the added value of IBM today is much less than that of personal computers. It didn't have to turn out that way.

5 Changing the Added Values

If you have little competition, your added value is assured. Then the strategic issue is whether and how to limit the added values of the other players in the game. We saw this played out with Nintendo. It established a virtuous circle that gave it a monopoly in 8-bit video games. For a time, there was no real threat from competitors. But the other players in Nintendo's Value Net – customers, suppliers, and complementors – still had claims on the pie. Nintendo's strategy limited the added values of all these other players.

Most of the time, there's plenty of competition – in which case the challenge isn't how to limit the added values of other players, it's how to have some added value of your own. Building added value is the hard work of basic business. You look for ways to improve quality at a low incremental cost, as TWA did with Comfort Class, or reduce costs without sacrificing too much in quality. Even better than making these intelligent trade-offs is finding what we call trade-ons – opportunities to improve quality and reduce costs at the same time.

Competitors play the same game. They work to make similarly smart trade-offs and trade-ons. This dynamic erodes your added value. To protect your added value, you need to create relationships with your customers and suppliers. Without a relationship, you could be selling a commodity. With a relationship, you're sure to be selling something unique – part of the package is you. The relationship provides a boost to your added value. In the presence of competition, it's often the key to making money.

American's AAdvantage frequent-flyer program is the model example of engineering a relationship. It creates loyalty by rewarding it. We suggested that every business should have a loyalty program and gave nine tips on how to be most effective at saying thank you to your customers.

Loyalty programs are an example of a strategy that continues to work even when imitated. Contrary to conventional wisdom, imitation can sometimes be healthy rather than harmful. The reason is that when strategies have a win-win component, that component gets amplified by imitation. More examples of healthy imitation are coming up.

Added value is the primary source of power in a game, but it's not the only source. Rules can alter the balance of power between the players. How rules do this is the subject of our next chapter.

6 Rules

> **When the rules of the game prove
> unsuitable for victory, the gentlemen
> of England change the rules.**
> – Harold Laski[1]

When we talk about changing the game, the first thing people usually think of is changing the rules. But if we ask what rules you might change or how you might go about changing them, the question often seems perplexing. After all, most of the rules businesspeople play by are well-established laws and customs. They have evolved to help ensure that trading practices are fair, that markets keep operating, and that contracts are honored. To step outside these rules would be to risk legal penalties or exclusion from the market.

But there are other rules of the game that it can make sense to change. Many of these rules are the ones found in contracts. Your contracts with customers and suppliers shape your transactions with those players in ways that extend far into the future. A single clause can tilt the balance of power heavily toward you or against you. By shaping your relations with customers and suppliers, these same contracts will also shape your relations with competitors. To be sure you're in a game where you'll make money, you have to make sure you've got the right rules in your contracts.

What all these more negotiable rules have in common is that they involve 'details.' Compared with changes in players or in added values, the possible changes in rules can seem like a small-scale matter. This makes them easy to ignore:

> **I want to know God's thoughts; the rest are details.**
> – Albert Einstein

But there's another way of looking at it:

> **God is in the details.**
> – Ludwig Mies van der Rohe

As we'll show in this chapter, relatively small changes in the rules of business can produce enormous changes in the outcomes. In other words, where business rules are concerned, the details are everything.

To illustrate how this works, we'll look at a variety of rules and analyze how each one affects the game. This requires imagining that a given rule is in effect, putting yourself in the other players' shoes, and playing out the game from all the different perspectives. With a better understanding of the rule's consequences, you can then decide whether you want to employ that rule or, if it's a rule that's already there, whether you want to change it.

There is no mechanism, or algorithm, for generating rules. It's a creative act. Still, you can get ideas for new rules from a number of sources. One approach is to find a rule that works in one context and consider whether it would work in a different one. Take a rule used with your customers and bring it to negotiations with your suppliers. Or take a rule you see used to good effect in other businesses and apply it to your own business. The collection of rules discussed in this chapter should serve as a useful source of ideas.

1 Contracts with Customers

You and your customers are partners in creating value, but it's not all cooperation. There's an inevitable battle when it comes to dividing the pie. When your customers press you for price concessions, that's competition, not cooperation.

In the Players chapter, we talked about bringing in more customers as a way of shifting the balance of power in your favor. You may even want to pay people to play, as LIN Broadcasting did with BellSouth. In the Added Values chapter, we saw how restricting supply worked to limit the added values of customers. That's one reason why Nintendo and DeBeers have been so successful.

In this section, we look at how to use rules to change the game with your customers. Because rules alter the balance of power, you can use them to restructure negotiations to your benefit. Of course, your customers will also be trying to change the rules to put themselves in a stronger position. The battle to establish the rules is the battle before the battle. The question of whose rules will rule is something we'll come back to at the end of the chapter. But the immediate goal is to develop a clear understanding of how rules change the game.

We'll begin with the 'most-favored-customer clause.' Such clauses are widely used, but their full implications are not widely appreciated.

Most-favored-customer clauses

A most-favored-customer clause (MFC) is a contractual arrangement between company and customer that guarantees the customer the best price the company gives to anyone. The MFC prevents a company from treating different customers differently in negotiations. Other common names for this rule are 'most-favored-nation clause' and 'best-price provision.'[2]

MFCs are very common in business-to-business contracts. We've seen them in businesses ranging from aspartame to aluminum cans to auto parts to fiber optics to building control equipment. Customers often like the price insurance feature. With an MFC in place, the customer is guaranteed never to be at a cost disadvantage vis-à-vis any competitors who buy from the same supplier.

MFCs sound like a good deal for your customers, but what do they do for you? To find out, let's revisit the Card Game from the Game Theory chapter.

The Card Game Continues Adam and twenty-six of his MBA students are again playing a card game. As before, Adam holds twenty-six black cards, and each student holds a red card. The dean is still agreeing to pay $100 to anyone who turns in a pair of cards, one black and one red. Once again it's a negotiation game between Adam and the students, like the first version of the Card Game in the Game Theory chapter.

This time, though, there's a new feature. Just before the game begins, one student, Tarun, announces that he has to leave for a job interview. No problem. Adam promises Tarun that he'll get the best deal given to any of the other students – Adam gives Tarun an MFC. Tarun figures he now has the best deal of all, and he doesn't even have to work for it. Off he goes.

Even though Tarun will get the best deal, he might be surprised at what that deal turns out to be. Adam's contract with Tarun is going to make Adam a much tougher negotiator with each of the other students. Let's look at the first of these negotiations. If the student pushes for an extra dollar, that costs Adam a dollar right now. But the concession also costs Adam another dollar, since he has to give the same price to Tarun. Whenever Adam makes a new concession of a dollar to the student, it costs him two dollars.[3] The bargaining is no longer symmetric. Adam will push twice as hard as the student does. He'll be a braver, more aggressive negotiator. Consequently, we expect Adam will end up with more than half of the $100.

It's more or less the same story with every one of the twenty-five students with whom Adam negotiates. Even though Tarun is not in the room, his presence is felt in every negotiation. All the students will get a worse deal – and that includes Tarun. When he returns from his interview, he'll be quite disappointed to see how the Card Game played out.

MFCs are counterintuitive in their effects. The natural guess is that customers do better with the protection of an MFC. And they would – if MFCs didn't change the game. But they do change the game.

If you were surprised by this result, you aren't alone. On more than one occasion, the United States Congress has chosen to play Tarun's role. It has left negotiations to others and been content to take the best price. Did the government do any better than Tarun did in the Card Game? Let's find out.

Favoring the House One of the many fringe benefits that members of Congress enjoy is that they get to make the rules. And others have to play by them. Needless to say, some of the rules nearest to Congress's heart are those that concern elections and campaign expenses.

Back in 1971 members of Congress figured that they could spend less time fund-raising if they could find a way to lower the cost of election campaigning.[4] A particularly expensive campaign item was television spots. Thus, Congress passed the Federal Election Campaign Act, which requires television broadcasters to automatically give candidates a rate for campaign spots equal to the lowest rate given to any commercial customer. The politicians effectively voted themselves an MFC for buying airtime.

The law didn't have quite the desired effect. Knowing that, in an election year, politicians are going to purchase significant chunks of airtime, the networks want to get as much as they can for these campaign spots. So with an election coming up, how will a network respond when a commercial customer, such as a Procter & Gamble, comes to negotiate the rate for an ad spot? It'll be very tough on price. Giving P&G a price concession, even to fill up slack airtime, is extremely costly, since any discount to P&G must be extended to all the politicians buying spots. The network would likely end up losing more from the lower price paid by all the politicians than it would gain from getting some extra P&G business. The bottom line: P&G doesn't get the discount.

One result of the law was that the networks ended up making more money than before. They used the fact that the politicians had an MFC to restrain themselves from giving price breaks to all their other customers. For these customers – Procter & Gamble and others – the law was like a hidden tax. In fact, the politicians may have succeeded in taxing themselves. Even though they got the best price, the best-price provision led the networks to charge everyone else more. Thus, that best price could well have been higher than the price they would otherwise have had to pay. Passing the law may have actually made the politicians worse off.

This wasn't the only time Congress shot itself in the foot this way. In 1990, as part of the Omnibus Budget Reconciliation Act, Congress reformed Medicaid reimbursement. Looking for ways to control drug prices, Congress was frustrated by the fact that some of the large HMOs had gotten lower prices than the government itself had. So Congress changed the game. It passed legislation specifying new rules for how Medicaid would pay for any branded drugs. Henceforth, Medicaid would pay 88 percent of the average wholesale price, or the best price given to anyone in the retail pharmaceutical trade, whichever was lower.

How did the government fare in the new game? Not as well as it expected. Look at the new game from the perspective of a drug manufacturer. With the new law, it's not worthwhile for you to offer anyone a price below 88 percent of the average price. If you did, you'd have to extend this low price to the government, and that would surely cost you more than you'd get from the extra business.

Dale Kramer, director of drug purchasing for Kaiser Permanente, America's largest HMO, described the phenomenon: 'In the past we'd offer a [drug] manufacturer 90 percent of our business, maybe $10 million in additional business, and get really low prices. Now, no one wants to go below the Medicaid floor.'[5]

That's not all. Since no one gets a price below 88 percent of the average price, the average doesn't stay put. As the prices at the low end vanish, the average price rises. Now a drug manufacturer isn't going to offer anyone a price below 88 percent of the new, higher, average price. The reason is the same as before. And the result is that the average price in the market rises again. On it goes. It's hard to say where all this ends up, but it's possible that 88 percent of the eventual average market price will be higher than the original average.

For the government, insisting on paying only 88 percent of the average price in the market was quite sensible. The mistake was awarding itself an MFC at the same time. That gave the drug manufacturers an incentive to raise prices, and up the average went. The government got hit with the higher prices – and so, for that matter, did everyone else. What the government should have done was find a way to give the drug manufacturers an incentive to price low, not high. Then the average price would have fallen, and the government would have been able to cut its drug bill even further.

Stanford Business School professor Fiona Scott Morton estimated the effect of the 1990 change in rules on drug prices.[6] She concluded that prices went up by an average of 9 percent for branded drugs on-patent and 5 percent for branded drugs off-patent. Even the prices of generic drugs rose, despite the fact that generics weren't covered by the best-price provision. Generic prices went up by an average of 2 percent, riding the coattails of the rise in the price of branded drugs.

The way the new game played out may have been a surprise to the government, but it wasn't to the drug makers. They had it right all along. Drug makers didn't exactly object to having to give the government the best price. Roy Vagelos, CEO of Merck, explained: 'Our concept of best-price for Medicaid — now incorporated into federal law — was supported by our longstanding policy of avoiding deep discounts.'[7]

In voting itself MFCs, Congress ended up helping out the television networks and drug manufacturers at the expense of advertisers and HMOs. Congress didn't do itself any great favor, either.

MFCs change the game. When your customers have MFCs, you're more able to withstand pressure to lower price. There's a common ritual in negotiations with customers over price. You say to the customer: 'I'd love to give you a better price, but I can't afford to.' The customer responds: 'You can't afford not to. If you don't, I won't buy from you.' Often you lose the argument. But if your other customers have MFCs, your argument becomes a lot more convincing. You can point out that a price concession to one will, by necessity, become a price concession to all. And that's something you really can't afford. You can just say no.

MFCs are an instance of 'strategic inflexibility.' People often think that having more flexibility is one of those universally good things. It isn't. Sometimes you have more power when your hands are tied.

This was the strategy employed by Hernán Cortés, the Spanish conquistador, upon his arrival in Mexico. His troops were vastly outnumbered. Fearing defeat, many of them wanted to turn back. To strengthen his men's resolve, Cortés had his ships beached and then taken apart.[8] With the option of retreat eliminated, Cortés's soldiers fought their way inland. By the time they reached the Aztec capital, Montezuma was ready to surrender without a battle. Cortés's strategy, though it reduced his options, strengthened his hand.

Giving a customer an MFC works the same way. You've reduced your options, since you're now forced to give the customer the best price you give anyone else. This strategic inflexibility is just what you want. It ties your hands when you negotiate, enabling you to stand up to your customers.

If MFCs help the seller capture more of the pie, why do customers go along with them — even press for them? One reason is that some customers simply don't get it. They don't realize how an MFC changes the game. That's hardly surprising, considering that the way MFCs work is rather subtle, even counterintuitive. A second reason is that some customers recognize they are poor – or, at best, average – negotiators. These customers do better taking the lowest price anyone else negotiates, even if MFCs lead to higher prices overall. Third, MFCs don't always lead to higher prices. There's always the chance that someone will turn out to be a really tough customer. If the seller is forced to give this customer very generous terms to get the deal done, he'll have to go back and give the same terms to everyone else with an MFC. These other customers will then get price breaks they might have been unable to get for themselves.

Corporate customers have another reason to accept, even desire, an MFC. A corporate customer may be less concerned with the absolute level of prices than he is with finding himself at a cost disadvantage vis-à-vis his competitors. If so, having an MFC makes a great deal of sense. It's effectively an insurance policy that guarantees the customer cost parity with anyone else who buys from the same supplier. The premium on the policy is the likely increase in the overall level of prices.

Customers' purchasing agents are often the keenest devotees of MFCs. The last thing a purchasing agent wants is for a competitor's purchasing agent to get a better price. It will look as if the agent isn't doing his job – and that's a good way to get fired. An MFC solves the purchasing agent's problem. He may not realize that MFCs lead to higher prices overall, or, even if he does realize, he may not care too much so long as he gets the best price in the market. What that price is, well, that's someone else's department. The situation is reminiscent of this verse from Tom Lehrer:

> 'Once the rockets are up, who cares where they come down?
> That's not my department,' says Wernher von Braun.

We've now looked at how MFCs benefit sellers and why customers go along with them. That leaves one more aspect to discuss: how MFCs, once in place, change the way customers negotiate.

MFCs reduce customers' incentive to negotiate. We saw what happened when the government voted itself an MFC and then sat back and let others do all the negotiating. Most customers who take MFCs play a more active role. They don't give up the chance to negotiate for themselves. Even so, the typical customer with an MFC won't push as hard in negotiating with the seller. That makes sense. The customer might as well let others do some of the hard work, secure in the knowledge that he'll benefit from any price breaks that they extract from the seller. Of course, if everyone lets everyone else do the hard work, the hard work never gets done.

This slacking-off effect is especially strong if what the customer really cares about is cost parity vis-à-vis his competitors. With an MFC, the customer is guaranteed that no one else can get a better price than he can, even if he sits back and does no negotiating at all. Should the customer work hard to try to extract a low price from the seller, in the hopes of opening up a cost advantage over his rivals? Probably not. Chances are his rivals also have MFCs from the seller. If the customer gets a good price, so will everyone else. He'll have put in a lot of effort and not gained any advantage. Why bother?

In sum, MFCs turn sellers into tigers and customers into pussycats. So who do you think gets the lion's share of the pie?

What ultimately makes MFCs so effective in changing the game is the subtle way in which they enable a seller to take control, almost in a backdoor fashion. By giving out an MFC to one customer, you change the game for everyone else. That's because when you negotiate with someone, much more important than whether that customer has an MFC is whether your other customers have MFCs — that's what makes you a tiger. The customer may not like that fact, but there's nothing he can do about it. He has no control over whether your other customers have MFCs; all he controls is whether he has an MFC. And if other customers have MFCs, thereby putting him in a weaker position, he might as well accept an MFC and get the insurance protection it offers. When he does so, that reinforces everyone else's incentive to take an MFC.

Despite the great advantages of MFCs, sellers shouldn't think of them as a panacea. One drawback of giving out MFCs is that doing so makes it harder to keep your customers. Suppose a competitor attempts to steal a customer away with a low price. To keep the customer, you'll probably have to meet the competitor's price. But matching the low price sets a precedent among your other customers, who may then expect the same break. If those other customers have MFCs, they'll not only expect the same break — they'll get it. In that case, holding on to the first customer may well be too expensive. You have to let him go.

That's exactly what your rival may be counting on. Knowing that your customers have MFCs, he might be all the more tempted to go after them in the first place. So if your concern is losing customers to a rival, giving them MFCs may be a bad idea.

A second drawback is that it becomes more expensive to go after a rival's customer with a low price. You have to give the same deal to all your existing customers, and that's unlikely to be worth it.

Of course, for these same two reasons, it's to your benefit when a rival gives MFCs to his customers, but that's not under your control.

MOST-FAVORED-CUSTOMER CLAUSE:
THE SELLER'S PERSPECTIVE

PROS

1. Makes you a tougher negotiator.
2. Reduces your customers' incentive to bargain.

CONS

1. Makes it easier for a rival to target one of your customers.
2. Makes it harder for you to target one of your rival's customers.

MOST-FAVORED-CUSTOMER CLAUSE:
THE CUSTOMER'S PERSPECTIVE

PROS

1. Allows you to benefit from any better deal subsequently offered to other customers.
2. Ensures that you're not at a cost disadvantage relative to rivals.
3. Eliminates the risk of looking bad if other customers strike better deals.

CON

1. When others have MFCs, it's harder for you to get a 'special' deal.

We've seen that the main effect of MFCs on the seller-customer relationship is to shift the balance of power in favor of the seller. We've also seen that giving out MFCs isn't without risk; doing so makes you more vulnerable to competition from rivals. If poaching of your customers is your main concern, you want to look for a different type of rule to change the game.

Meet-the-competition clauses

A rival comes after your customers. What can you do? One way to make it harder for him to steal your customers is to employ a meet-the-competition clause (MCC). An MCC is a contractual arrangement between company and customer that gives the company an option to retain the customer's business by meeting any rival bids. Of course, an MCC doesn't force you to meet the competition. It simply rewards you, if you do so, with the assurance of the customer's continued business.

There are several different names for a meet-the-competition clause; sometimes it's called a 'last-look provision'; other times it goes under the name 'meet-or-release clause.' These are just different names for an MCC. Whatever the name, MCCs are most often found in commodity businesses.

I've Met the Competition and. . . To understand how MCCs work, put yourself in the position of a typical commodity producer. You don't have much bargaining power and you regularly get forced down on price. That seems ironic because your product is an essential input. The problem, of course, is that you're not the only producer.

There are some saving graces. The commodity is expensive to transport, which gives you some added value wherever you're the one best located to serve the customer. You also get some added value from your reliability, reputation, service, and technology. Still, your added value is small in relation to the total pie. That's almost the definition of a commodity business. The question is: how can you improve your bargaining position?

Having MCCs with your customers can help you sustain a higher price. Normally, an elevated price would invite your competitors to undercut you. If you have an MCC in place, however, a rival can't come in and take away your customers simply by undercutting your price. If he tried, you could then come back with a lower price and keep the business. The back-and-forth could go on until price fell to variable cost, but at that point it wouldn't be worth the rival's effort to steal your customer. The only one to benefit would be the customer, who'd end up with even more of the pie.

You can see the strength of your position by stepping into your rival's shoes. He runs a risk anytime he cuts price to go after your business. Remember the Eight Hidden Costs of Bidding from the Players chapter? We recast them below as they apply to a rival's bidding for one of your customers:

EIGHT HIDDEN COSTS OF BIDDING FROM
A RIVAL'S PERSPECTIVE

1. He's unlikely to succeed – there are better uses of his time.
2. When he wins the business, the price is often so low he loses money.
3. You can retaliate – he ends up trading high-margin for low-margin customers.
4. Win or lose, he helps establish a lower price – his existing customers will then want a better deal.
5. He'll set a bad precedent – new customers will use the low price as a benchmark.
6. You will also use the low price he helped create as a benchmark.
7. It doesn't help him to give his customers' competitors a better cost position.
8. He shouldn't destroy your glass house – if you're a little vulnerable, you'll be less likely to go after his accounts.

Despite this list of reasons not to go after your customers, a rival might still try, in hope of gaining some new business. But when you have MCCs, that justification is much weaker. Now going after your customer has all the old downsides and even less upside. The rival would do better to make sure that his existing customers are happy.

Putting in MCCs changes the game in a way that's clearly a win for you. As for rivals, while it's true that they have less ability to take market share from you, there is – perhaps surprisingly – a win-win element here, too. Your higher prices set a good precedent: they give rivals some room to raise price to their own customers. What's more, you're less likely to go after their customers because, with your higher profits, you have more to lose. That's the glass-house effect again.

As for customers, why do they go along with MCCs? It may be the norm in their industry. Even without a formal MCC, it is generally accepted that customers don't leave their current suppliers without giving them a last chance to bid. In some cases, it may be that customers' purchasing agents are focused on the short term. In return for a price break today, they're willing to accept less bargaining power tomorrow. In other cases, it may be that customers don't thoroughly understand the rule's implications.

Whatever the reason, MCCs do offer some benefits to customers. This is because MCCs assure producers of a long-term relationship with customers, even in the absence of long-term contracts. With this assurance, producers are more willing to invest in serving their customers better and more likely to share technology and ideas. This partnership orientation can lead to a long-term win for the customer.

If you're a seller, remember that asking for an MCC is a way of getting paid to play. If a customer wants you to bid, but won't pay you to play in cash, ask for an MCC. If one of your existing customers has solicited fresh bids for his business, forcing you to come down on price, ask for an MCC in the new contract. The customer may consider this a small concession to make in light of the price concession you've just made. Since having the MCC will take the guesswork out of your future pricing, you won't have to lower your price preemptively. If another seller underbids you, you'll now have the chance to respond. And, as we've pointed out, others will now have much less incentive to underbid you.

Healthy Imitation Far from being undermined by imitation, a meet-the-competition clause is actually enhanced by it. An individual producer benefits unilaterally from inclusion of MCCs in its contracts with buyers. There's an added benefit when other producers also put in MCCs. Their MCCs allow them to push prices up some more, so they now have even more to lose from starting a share war. As MCCs become more widespread in an industry, everyone has less prospect of gaining share. With even more at risk and even less to gain, producers refrain from going after one another's customers. Now everyone is firmly ensconced in a glass house.

Without things like MCCs, competition in the marketplace is a bit of a free-for-all. One seller can come in and try to take away another's customer. In this game, there aren't any rules of engagement. The customer doesn't have to go to his incumbent supplier and give it an opportunity to match. Even if the incumbent gets that opportunity, the customer may not tell him what number he has to beat. In this case, the incumbent may end up bidding lower than he'd have had to in order to keep the business.

An MCC changes the game by creating rules of engagement. A rival can still make a bid, but now the incumbent has the final move. The customer can't switch without first revealing the rival's bid and giving the incumbent a chance to match it. If the incumbent matches the rival's bid, he's guaranteed to keep the customer. Knowing that the incumbent will get the last look, the rival has much less incentive to play the game in the first place. That's why MCCs put an incumbent seller in a powerful position.

As with most-favored-customer clauses, meet-the-competition clauses aren't a panacea for sellers. They create a vulnerability if you face a rival whose main objective in life appears to be hurting you rather than doing well for himself. Normally, if a rival comes in with a low price, he'd better be prepared to deliver. If you have an MCC, however, your rival can make a low bid, fully expecting that you'll match and he won't have to make good on his offer. He can lower your profits without having to put himself on the line. We don't think this strategy would be in your rival's self-interest, but you can't assume that your rivals will always see their self-interest the way you do.

MEET-THE-COMPETITION CLAUSE

PROS
1. Reduces the incentive for competitors to bid.
2. Takes the guesswork out of bidding – you know what bid you have to beat.
3. Lets you decide whether to keep the customer.

CON
1. Allows competitors to bid without having to deliver.

There's a buyer's counterpart to an MCC. A buyer would like a guarantee that the seller will sell to him provided he matches the highest price anyone else offers the seller. Used this way, the rule is typically called a right of first refusal. But conceptually, it's exactly the same as an MCC. Both entitle the person to a last look. That's the key. And, given what you've just seen with MCCs, you won't be surprised to learn that a right of first refusal puts the buyer in a powerful position.

Blocked Bidders In January 1994 the Miami Dolphins football team was sold for $138 million to H. Wayne Huizenga, founder of Block-buster Video. A pretty good buy – almost a steal. By contrast, the New England Patriots were sold for $160 million around the same time. The price the Miami Dolphins fetched was even below what the NFL charges people for the right to create an expansion team – a new team with no track record, no coach, nothing. The Miami Dolphins was a team with five American Conference titles, two Super Bowl victories, pro football's best record since 1970, and legendary coach Don Shula, to boot.

Why was the price so low? Partly, it was a distress sale. The Dolphins had done brilliantly under their longtime owner, Joe Robbie. But when Robbie died in 1990, the team was passed to his nine children. Conflict among the children, together with a $30-million estate tax liability, soon forced a sale. Even though it was a forced sale, that doesn't explain why buyers weren't lining up.

The key to the explanation is in the contract that the Robbie children had earlier given Huizenga. Following the death of their father, the Robbies sold Huizenga a 15 percent stake in the Dolphins and also gave him a right of first refusal on any future sale of the team. Thus, the Robbie children couldn't sell the Dolphins without first giving Huizenga an opportunity to match the best offer.

Put yourself in the shoes of a prospective bidder for the Dolphins. You invest time, effort, and money lining up financing and hiring investment bankers to do valuations. Will you be able to outbid Huizenga? Doubtful. If it makes sense for you to acquire the Dolphins at a certain price, it makes sense for Huizenga, too. And he gets the right to match your bid and get the team. That's the best-case scenario. The worst-case is that you actually win the bidding. If Huizenga doesn't match your bid, that's strong evidence you've overpaid.

Actually, a bidder was in an even worse position than this. Huizenga owned half the stadium where the Dolphins played and also owned the Florida Marlins baseball team, which shared the stadium with the Dolphins. With all these synergies, it's hard to imagine that acquiring the Dolphins could have been as valuable to anyone else as it was to Huizenga. And someone who outbid Huizenga would have had to negotiate with him over the use of the stadium. Bottom line: entering the bidding for the Dolphins was a· losing proposition for anyone but Huizenga.

As it happened, very few people took a serious look at the Dolphins, and there were only two outside bids. One had so many conditions attached that the Robbie children rejected it without even bringing it to Huizenga. The other was the $138-million bid that Huizenga matched when he bought the team. The Wall Street Journal quoted one investment banker's view of the deal: 'If you let somebody take control of the process on the buy side, you can't get up a head of steam from other outside bidders.'[9]

What should the Robbie children have done? They shouldn't have given a right-of-first-refusal provision, or at least not without getting paid handsomely for doing so. Even after Huizenga had his right-of-first-refusal provision, they still could have done better. We saw the solution in the Players chapter. LIN Broadcasting was in a very weak position after Craig McCaw made his hostile takeover bid. That's why LIN paid BellSouth $54 million to come into the auction. For a lot less money, the Robbie children could have – and should have – paid other people to enter their auction. But they didn't.

2 Contracts with Suppliers

You and your suppliers, just like you and your customers, are partners in creating value. But here, too, it's not all cooperation. When your suppliers try to raise prices, that's competition.

In the Players chapter, we talked about bringing in more suppliers as a way to shift the balance of power in your favor. Gainesville Regional Utility brought in Norfolk Southern railroad as a way to counter the power of its incumbent supplier, CSX railroad. American Express created a buying coalition to bring in more health-care suppliers. In the Added Values chapter, we mentioned how the NFL has restricted the number of teams and their roster size, in part as a way of limiting the added values of football players.

Here, we'll look at how rules can be used to change the game with your suppliers. The Value Net suggests that for every rule toward customers, there's a symmetric counterpart with respect to suppliers. So far we've looked at two rules toward customers: a most-favored-customer clause (MFC) and a meet-the-competition clause (MCC). Each of these can be turned around.

The supplier-side analogue to an MFC is guaranteeing a supplier that you'll pay him at least as much as you pay any other supplier of the same resource. Call this a most-favored-supplier clause. Such clauses – either explicit or implicit – are sometimes seen in compensation contracts. Promising to pay someone as much as you pay someone else sounds like a generous offer.[10] In fact, the main effect is to put you in a better position to hold the line on all salaries. The effect parallels that of an MFC.

An MCC is essentially the same contract with suppliers as with customers: it gives you a last look. The only difference is that this time you get a right to buy as opposed to a right to sell. Formally, an MCC with a supplier is a contractual arrangement between company and supplier that requires the supplier to continue supplying the company provided that the company agrees to match the best price anyone else offers the supplier for its resources. Just as customers typically pay more when they've granted an MCC, suppliers typically get paid less when they've granted an MCC.

In certain professional sports – notably, basketball and hockey – team owners have an MCC in some of their contracts with athletes. This ensures that they won't lose an essential player without having the chance to match any competing bid. Of course, anticipating that the current owner will likely match any bid, rival teams have less incentive to bid for a player in the first place.[11] The effect is to reduce the competition for athletes below what it would otherwise be.

In principle, games with suppliers should be exactly parallel to games with customers. But, in practice, the rules in these games are sometimes different. Most-favored-customer clauses and meet-the-competition clauses with customers are much more common than their supplier-side analogues. Our next rule, the take-or-pay contract, is really only seen on the supplier side.

Take-or-pay contracts

A take-or-pay contract is a rule structuring negotiations between companies and their suppliers. With this kind of contract, you either take the product from the supplier or you pay the supplier a penalty. For any product you *take*, you agree to pay the supplier a certain price, say $50 a ton. Furthermore, up to an agreed-upon ceiling, you have to *pay* the supplier even for product you don't take. Naturally, this 'penalty' price is lower, say $40 a ton. Thus, if the ceiling is 1,000 tons and you take 900 tons, you pay $50 a ton on the 900 tons you take and a penalty of $40 a ton on the 100 tons you don't use.

Take-or-pay contracts are often seen in agreements to purchase commodity inputs, electricity, even cable programming. Suppliers in these businesses face large fixed costs relative to variable costs. In some cases, it's also impractical for the supplier to store the output. As a buyer, you have enormous power to beat down price. For protection, the suppliers seek take-or-pay contracts – the larger the better. Of course, you recognize the exposure that comes from signing a take-or-pay contract, and you won't agree to take more than you expect to use. That's why take-or-pay contracts often end up being for amounts close to predicted demand.

By agreeing to a take-or-pay contract, you clearly help your supplier. The contract enables the supplier to plan production better. It also makes him less vulnerable to being held hostage by you later on. In return for providing him with this security, the supplier will likely reward you with a lower price.

But that's not the only benefit you can get from entering into a take-or-pay contract with your supplier. Take-or-pay contracts can change the game in another way, too. They can affect the pricing dynamics in your industry by reducing the incentive of a rival to go after your customer base.

If a rival goes after one of your customers, he risks retaliation – that's one of the Eight Hidden Costs of Bidding. If you have a take-or-pay contract with your supplier, that retaliation becomes a near certainty. To understand why, imagine that you expect to use 1,000 tons of input and have a take-or-pay contract for that same amount. If a rival steals one of your customers, you'll need less input from your supplier. Suppose that, instead of the 1,000 tons you're currently using, you now need only 900 tons of input. Because you're still paying for the next 100 tons of input, even if you're no longer using it, you'll surely look around for some new business to replace the business you've just lost. Your variable costs are now a mere $10 a ton – the difference between the $50 'take' price and the $40-a-ton 'pay' price. At this cost level, you can hardly afford not to replace the business you've lost, and your rival's customer base is the obvious place to look.

With a take-or-pay contract, you're paying part of your input costs up front. You've turned some of your variable costs into fixed costs. If a rival takes one of your customers, the economics virtually force you to take one of his in return. Thus, take-or-pay contracts have a deterrence effect. Smart rivals will recognize the greater likelihood of retaliation and refrain from going after your customers in the first place.

Take-or-pay contracts can help stabilize market share in your industry. That's more good news for your supplier. If you don't have to fight to keep every customer, you'll make more money. And if you're making more money, you're less likely to fight with your supplier over cost.

There is a caveat. It's always possible that someone will act rashly and go after your market share, even though you have a take-or-pay contract. You're forced to retaliate, and that in turn could trigger a sequence of tit-for-tat responses that escalate into a full-scale price war. The war may be particularly brutal, since a part of your costs are already sunk. The deterrence effect of a take-or-pay agreement is not unlike nuclear deterrence: you hope it works, because the cost if it doesn't is extremely high.

```
┌─────────────────────────────────────────────────────┐
│                  TAKE-OR-PAY CONTRACT                │
│                                                      │
│                         PROS                         │
│   1. Reduces risk to your supplier, in return for    │
│      which you can ask to pay less.                  │
│   2. Reduces a rival's incentive to come after your  │
│      customers by making retaliation a near          │
│      certainty.                                      │
│                                                      │
│                         CON                          │
│   1. Increases severity of price war if deterrence   │
│      fails.                                          │
└─────────────────────────────────────────────────────┘
```

3 Mass-Market Rules

So far we've been looking at rules in business-to-business settings. When businesses deal with each other, buyers and sellers negotiate not just over price but over the rules of the game as well. For example, you may want to have a meet-the-competition clause, but the customer may balk. Since you don't have the power to impose rules unilaterally, what the rules will be is itself subject to negotiation.

Mass consumer markets are very different. Sellers don't negotiate. And since sellers don't negotiate, buyers can't negotiate. As a seller, you have the power to unilaterally lay down some rules of the game. One rule is that you get to name the price of the item you're selling. If the customer wants your product, he has to pay your price. That's the way it is in supermarkets, gas stations, restaurants, department stores – in fact, in almost every part of the retail sector.

This state of affairs is something people take for granted. Still, it's interesting to ask why it's the case. Once again, the Card Game proves illuminating.

Just for a moment, imagine that Adam has one hundred decks of cards and is playing the Card Game simultaneously with 2,600 students.[12] Paralleling the analysis of the very first version of the Card Game, you'd say that Adam and the students will split the $100 prizes equally. Nothing has really changed.

Or has it? The fact that it's a large-numbers game makes a big difference. Adam can very plausibly refuse to negotiate with each student individually. There just isn't the time. Instead, he can announce a price and require students to take it or leave it.

The large numbers effectively turn the situation into the ultimatum version of the Card Game. As we saw in the Game Theory chapter, that puts Adam in a much stronger position. He gets to pick the price. He can offer $10 for a red card and expect the students to accept his offer. They'll do so because they know they're not going to get an opportunity to make a counteroffer.

If you're selling to a mass market, then, just like Adam you can refuse to negotiate. Instead, you get to set a price. What other rules might you want to impose?

At first glance, there'd be nothing to gain by imposing a most-favored-customer clause in a consumer market. If you have the power to set price, there's no issue of customers pushing for special deals. But, in fact, there is still a reason to consider an MFC. Though customers may be unable to negotiate, they can postpone buying. They may do just that if they're not convinced that you'll hold firm to your price. The more customers that wait, the more pressure you'll feel to lower price. The customers' belief that price will fall becomes a self-fulfilling prophecy. Granting an MFC helps you get out of this trap.

January Sales In 1990 Chrysler used a variant of an MFC to change the game of selling cars. The way car buyers played the game is that they would wait for end-of-year rebates, and, by waiting, leave the dealers with large inventories, which would then force the manufacturer to offer these rebates.

Chrysler wanted to convince customers that there would be no gain from waiting. Mere words wouldn't have been credible. What Chrysler did, as can be seen from the advertisement reproduced on the next page, was promise people who bought a car in January that if it offered a larger rebate later in the year, it would go back to them and make up the difference.

Chrysler's guarantee had two effects. Customers now had no incentive to wait, so end-year inventories were smaller and end-of-year rebates weren't as necessary. Moreover, Chrysler was now less tempted to offer rebates to clear out whatever end-year inventory did remain. It would be too costly to go back and make refunds to everyone who bought in January.

Many retail stores offer thirty- or sixty-day price protection. The motivation is similar to Chrysler's. Every Day Low Pricing (EDLP) policies are another rule that retailers use to convince customers not to wait for sales, and, in this way, stop themselves from being forced into having a sale.

How about the use of a meet-the-competition clause in a consumer market? Here, there's a catch. You can offer to meet the competition, but you can't force consumers to buy from you if you do. Most of the time you don't have a regular contract with your customers. When you do, it's not the kind that will tie them down. Try to imagine a potential car buyer signing a contract that gives Chrysler the option to match any offer made by Ford. Not very likely. Why would anyone ever sign such a contract?

So in a mass-consumer market, you can't quite make any rule you want. You can only make rules concerning what you do, not what the customer does. If you can't have an MCC, can you achieve essentially the same result by simply announcing your intention to match the best price in the market? Not entirely. A guarantee to match any competitor's price is not the same as an MCC. Consumers don't have to give you the option to match.

Some retailers do use a best-price guarantee to good effect. British retail chain John Lewis has as its motto 'Never Knowingly Undersold.' It is quick to reduce its price if it discovers that the same item is available for less elsewhere, and thus has a well-earned reputation for having the lowest price. The result is a very loyal customer base.[13]

A best-price guarantee doesn't work as well when what you sell is more idiosyncratic. Suppose that Chrysler tried to offer such a guarantee – say, on its Neon model. What's the comparable Ford, GM, Toyota, or Hyundai car? With respect to which models, with which options, would the guarantee apply? However Chrysler framed its guarantee, it would expose itself to considerable risk. If for some reason – a move in exchange rates, say – the price of a comparison model fell, Chrysler could be forced to match a price that would cause it to lose money.

Since you can't have an MCC, and giving a best-price guarantee

isn't always practical or wise, how else can you try to hold on to your customers in a mass consumer market? There's always the option of keeping your customers happy by simply charging them a low price. There are two immediate problems with this strategy. First, you've just lowered your profits. Second, establishing a low price is an offensive as well as a defensive move. Your low price will attract some of your rivals' customers away, and then your rivals will be forced to respond by lowering their prices in order to preserve their own customer bases. You're back to where you started, except prices are lower all around.

So what you'd really like to do is to charge a low price to your own customers without, at the same time, threatening your rivals' customer bases. If you could do that, your rivals wouldn't have to respond. This is a rather curious idea. Given the rules that normally govern business-to-consumer dealings, it doesn't even sound possible. But GM discovered a way of changing the rules. We'll see how with the story of the GM Card program.

General Motors Leads the Charge Flat demand and competition from foreign car manufacturers made the early 1990s a difficult period for the Big Three – General Motors, Ford, and Chrysler. In 1992 GM chalked up the largest annual loss in US corporate history, around $4.5 billion.

Part of the solution to GM's problem was to make better cars and make them more efficiently. GM hoped that the Saturn Corporation, an autonomous unit it had established in the mid-1980s to make small cars, would catalyze improvements throughout GM. Another promising sign was a late-1992 boardroom coup led by John Smale, a GM director and retired chairman of Procter & Gamble. This led to a major shake-up of GM management.

That still left the issue of how cars were sold. In response to flat demand and increased competition, each of the Big Three relied heavily on cash-back offers, dealer discounts, end-of-year rebates, and other incentive programs.

In September 1992 GM changed the game. It teamed up with Household Bank, a major issuer of co-branded credit cards and, together, they launched the GM Card under the MasterCard umbrella. Cardholders would earn credit equal to 5 percent of their charge volume, which could be applied to the purchase or lease of any new GM car or truck. Under the rules of the program, the credit was applied *after* the customer negotiated his or her best deal on the vehicle. These credits could become quite substantial: the limit was $500 a year, with a ceiling of $3,500 over seven years.

Visa USA President and CEO Robert Heller wasn't impressed. At the 1992 American Bankers Association bank-card convention, he joked that it wouldn't be long before pizza parlors joined AT&T and General Motors in offering cards. Within a year, Heller had been ousted from his position, and people were talking about the possibility of a McDonald's credit card.

The debut of the GM Card was supported by a marketing blitz that included 30 million direct-mail shots – referred to by GM as 'targeted tonnage' – 7 million telemarketing contacts, and extensive TV and print advertising. GM spent $120 million on the campaign. While small in relation to GM's total marketing costs, this figure was unprecedented for a credit-card launch.

The GM Card rollout was the most successful ever in the credit-card business. After only twenty-eight days, there were 1 million accounts. The previous record holder, AT&T's Universal Card, had taken seventy-eight days to reach that number. In less than two months, there were over 2 million GM Card accounts, and card balances topped $500 million.[14] At the one-year mark, the GM card program had 5 million accounts and $3.3 billion in outstanding balances. Two years out, there were 9 million accounts – and the program hasn't stopped growing.[15]

In the first year of the program, GM honored 55,000 card rebates. Through February 1994 there were 123,000 GM Card rebate redemptions worth a total of $40 million (an average of $325 per car). Projections suggest that as the program matures, some 25 percent of GM's nonfleet sales in North America will be to cardholders.

Targeted Rebates The GM Card is obviously a big deal, but how has it changed the game of selling cars? To see the effect, let's run through a simplified numerical example of how this kind of rebate program changes pricing dynamics. Suppose that GM and Ford are both initially charging $20,000 for their cars. Since prices are equal, the market divides on the basis of people's natural preference for GM or Ford.

Now suppose GM finds a way to direct a $2,000 rebate to its natural customer base and, at the same time, raises its list price to $21,000. At this point, GM is effectively charging two different prices: $19,000 to its own customer base and $21,000 to people who prefer Fords. Step into Ford's shoes. It can respond by lowering price to something under $19,000, in an attempt to steal GM's customers. Or it can go up to something close to $21,000, without danger of losing any of its natural customer base to GM. Ford will likely find the second option the more attractive. And if Ford does, in fact, go up to $21,000, that gives GM some breathing room. Now GM can raise price to something close to $23,000 – a net price of around $21,000 to rebate holders – without losing any of its customer base. At this point, GM and Ford are both ahead of the status quo. Ford might even raise price again, and then GM might, too, and so on.

The rebate program creates a win-win pricing dynamic between GM and Ford. How far could this process go? In practice, only so far. For one thing, there are other car manufacturers who might not have raised price. At some point, consumers might switch to them. In any case, the main benefit to GM is that Ford, or any other carmaker for that matter, is much less likely to start a price war.

The key to the effectiveness of the rebate program is that the rebates are targeted. It works only if the people who get rebates on GM cars are predominantly prospective GM buyers, and not prospective Ford buyers. Thus, a big practical challenge in implementing such a program is getting rebates to as many of your prospective customers as possible, without them also ending up in the hands of your rivals' prospective customers.

The GM Card solved this problem brilliantly. Recognizing that it's hard to go out and find all your prospective customers, GM turned the problem around. If it's hard to find your prospective customers, then let them find you. The people who are willing to build up a rebate via the GM Card are the ones who are planning to buy a GM car. People who intend to buy a Ford are much less likely than GM loyalists to use the GM Card. That's how a credit-card program helps you solve the problem of targeting rebates to your natural customer base.

Healthy Imitation II If the GM Card is as good as it sounds, then it was sure to get copied. And it was.

In February 1993, five months after the launch of the GM Card, Ford joined forces with Citibank, the largest bank-card issuer, to offer the Ford-Citibank Card under both the MasterCard and Visa umbrellas. The program allowed cardholders to accumulate up to $700 per year in rebates, with a cap of $3,500 over five years.

The Ford-Citibank Card was launched with a direct-mail solicitation to Citibank's 30 million existing cardholders as well as to owners of Ford cars. Promotional materials and applications were placed in five thousand Ford dealerships. Through the first nine months of 1993, a total of $4.6 million was spent advertising the Ford-Citibank Card. By April 1994 industry analysts were variously estimating the number of active Ford-Citibank cardholders at 1.3 million to 5 million. Some twenty thousand Ford customers redeemed Ford-Citibank card rebates in the first year of the program.

The GM Card was copied again in June 1994, when Volkswagen of America joined with MBNA Corporation, the leader in cobranded cards, to launch a credit-card rebate program. (Notable by its absence is Chrysler, which, as of early 1996, still didn't have a card program.)

Does all this imitation put a dent in the GM program? Not necessarily. If GM was hoping to use its program to take market share from Ford and others, then imitation was indeed bad news. With the Ford program now offering them a good deal, prospective Ford buyers had even less reason than before to acquire the GM Card.

But imitation also helps GM. As more card programs appear, car manufacturers are less tempted to cut price, because low prices are no longer as effective in attracting customers. People who have built up credits in someone else's program won't readily switch. Moreover, if a carmaker now raises price, it won't lose as many customers as before, because its own cardholders won't want to lose the rebate they've already earned. Carmakers now have more loyal customer bases. Overall, then, price cuts are less effective, and price rises less risky.[16] Not only is this true for manufacturers with a rebate program, but manufacturers without rebate programs will also find price cuts less effective. So there is less incentive to compete on price. The end result: more price stability in the industry. It's another case of healthy imitation.

Imitation has another positive effect: it gets customers off the sidelines. As more carmakers adopt rebate programs, it becomes more important for car buyers to choose sides. People who would not ordinarily display brand loyalty will recognize that they'll be in a disadvantaged position if they don't sign up for any of the programs. Even the more price-sensitive car buyers will have an incentive to lock themselves in. And the more customers that sign up for these programs, the greater the effect on pricing dynamics in the industry.

There's going to be more imitation down the road. Following the success of their programs in the United States, GM and Ford have launched credit-card programs in Canada and the United Kingdom, and Toyota has introduced a credit-card rebate program in Japan.

More Winners In addition to changing the pricing dynamics in the car industry, the GM Card program has many other benefits. Every month, GM encloses marketing material in Household's credit-card bill. It's better than free postage. Unlike so much junk mail, this envelope can't just be thrown away – there's a bill inside that has to be paid. When people open the envelope, out pops GM's insert.

GM also gets Household to share in the cost of the rebates, standard practice in the credit-card business. Issuers pay to offer rewards that make their cards more attractive. For example, First Chicago pays United Airlines roughly a penny per mile to give out frequent-flyer miles to its credit-card holders. Following the same principle, Household pays GM roughly 20 percent of the rebates, whether used or not.

What's good for GM is also good for Household. As a result of GM's marketing blitz, the 8-million-plus new accounts have propelled Household Bank from tenth to fifth place among credit-card issuers. The opportunity to earn a rebate on a GM car has helped lift annual charge volume on the GM Card to $5,200, two and a half times the national average. It's the first card people pull out of their wallet. Over 70 percent of receivables revolve on the GM Card, as compared with an industry average of 66 percent. Churn and delinquency rates for the GM Card are below industry average. Churn is reduced because it takes time to build up rebate credits, and, in the meantime, people are less likely to switch to another card. Even delinquency is lower, since the rebate can't be used if the credit-card account is past due.

Then there's the effect on pricing dynamics in the credit-card business. Now that Household, Citibank, and MBNA all have more loyal customers, there's less incentive for any credit-card issuer to compete on price. The result is greater price stability in the credit-card industry as well as the car industry.

If automakers and credit-card issuers are winners, does that mean that car buyers are losers? Not necessarily. Though they face higher prices, that's not the whole picture. Let's go back to the problem automakers faced in the early 1990s. Once a manufacturer has designed a car, spent money retooling the assembly line, and paid for a national ad campaign, it has incurred significant sunk costs. Throw in other manufacturers with similar cars, similar costs, and some excess capacity, and there's a big problem. Prices get pushed down to variable cost, and manufacturers don't earn back their investments. Competition sinks sunk costs.[17] This problem affects consumers as well as manufacturers. If manufacturers can't earn profits today, they don't invest, and consumers don't get better and cheaper cars tomorrow. So perhaps over the longer run, higher prices may lead to a win-win outcome for both the automakers and their customers.

Lessons in the Cards What has been the bottom-line effect of the GM Card program and its imitators? The fortunes of the automakers depend on many factors: the overall state of the economy, exchange rates, new-model introductions, their ability to forecast demand, and much more. So it's hard to disentangle the effect of the credit-card programs from everything else that has been going on. But it does seem that the programs are helping the automakers raise price by cutting back their traditional incentive programs. *BusinessWeek* noted the drop-off in incentives: 'Nearly all of the auto makers – and especially the Big Three domestic manufacturers who got consumers hooked on rebates in the past recession – are seeing big payoffs as they wean buyers away from cash-back offers and leasing subsidies.'[18]

On January 31, 1995, GM reported that its core North American auto operations (NAO) were back in the black for the first time since 1989. *BusinessWeek* explained how 'NAO has fattened its bottom line by curbing low-profit sales to rental-car fleets and by trimming . . . marketing incentives.'[19]

The story of the GM Card program provides an important lesson on pricing and the kind of rules you might want to govern it. People often think that the best strategy is to charge their own customers high prices and offer their rival's customers low prices. After all, while your own customers are willing to pay, your rival's customers need to be tempted by the offer of low prices. How does this strategy compare with the results of the GM Card?

The card appeals first and foremost to people planning to buy a GM car. Prospective Ford buyers are much less likely to have a GM Card. So what happens when GM scales back some of its other incentive programs to offset the cost of the credit-card rebates? The answer is that the effective price of a GM car – net of any rebates earned on the GM Card – is now higher to Ford loyalists than to prospective GM buyers. GM ends up pricing its cars higher to its rival's customers than to its own customers. This is the reverse of what people usually think is the best strategy.

But GM got it right! By raising price to prospective Ford buyers, GM gave Ford some breathing room to raise price – and that gave GM an opportunity to firm up price, and that gave Ford room to raise price some more, and so on. It's the win-win dynamic we talked about above. Contrast this with what happens when you price low to your rival's customers to tempt them away: you force your rival to respond by lowering price, which, in turn, puts you in danger of losing your own customers to your rival, and now you have to drop price to your own customers. You're worse off than when you began.

So the underlying principle is: treat your own customers better than your rival's customers. Companies seem to understand this idea in some contexts. For example, their efforts to improve products are often aimed toward strengthening ties with their existing customers, not toward luring away someone else's customers. But when it comes to pricing, companies often get it backward. They go after a rival's customers with low prices when, instead, they should focus on giving their own customers the best deal.

To prevent price wars, you want to charge your own customers low prices and offer higher prices to your rival's customers. Frequent-flyer programs accomplish just that. The price of a trip from New York to Chicago on American Airlines is the same to everyone – member of American's AAdvantage program or not. But the trip is more valuable to AAdvantage members than to nonmembers because it brings them that much closer to a free ticket to Hawaii. So by delivering its loyal customers more value for the same price, American is, in effect, charging them less than others for that trip from New York to Chicago. Who faces the higher effective price? Among others, all the people loyal to United Airlines: United loyalists won't value AAdvantage miles as much as American loyalists do. That means United now has some room to firm up its own prices without losing traffic, and that gives American some more breathing room, which in turn helps United, and so on. American and United enjoy the same win-win pricing dynamic as GM and Ford do.

There's clearly a close connection between a rebate program like the GM Card and the loyalty programs, such as AAdvantage, discussed in the previous chapter. That's to be expected. Both types of programs are rules – essentially unilateral contracts with customers, guaranteeing them a discount. Moreover, the way they affect industry pricing dynamics is the same. There's one major difference between the two types of programs. A loyalty program like AAdvantage rewards customers in kind rather than in cash and, in so doing, increases the size of the pie. By contrast, the GM Card rewards people in dollars and so doesn't increase GM's added value.[20] While it's better to have a program that rewards people in kind rather than cash, as we've just seen with the GM Card, cash can do nicely, too.

There's no reason why other businesses shouldn't follow GM's lead and change the rules by starting their own credit-card rebate programs. Rebate programs are most useful to businesses that sell big-ticket, infrequently purchased items. It isn't feasible for these businesses to reward customers in kind. Banks can't give out free mortgages, but they could offer rebates off mortgage points. Similarly, real-estate brokers can't give away houses, but they could offer rebates off their commissions. Indeed, anyone selling a big-ticket item might consider having a rebate program.

There's a final, ironic twist to the GM Card story. It appears that GM got it right for the wrong reason. In an interview, Hank Weed, managing director of the GM Card program, explained that the card was designed to help GM build share through the 'conquest' of prospective Ford buyers and others.[21] We believe that the conquest effect was modest, at best. The GM Card was always going to be most attractive to GM's own customer base and not that attractive to prospective Ford buyers. This was even more true once Ford started a card program of its own. At that point, prospective Ford buyers had virtually no reason to acquire a GM Card. Moreover, how could GM have expected its program not to be imitated? There was nothing proprietary about it. So it seems to us that the real effect of the GM Card was quite different: it improved the pricing dynamics in the car industry. In this, it was a runaway success.

REBATE PROGRAM

PROS
1. Allows you to charge your own customers low prices without threatening your rival's customer base.
2. Encourages customers – even price shoppers – to become loyal.

CONS
1. In rewarding loyalty in cash rather than kind, doesn't raise your added value.
2. Is ineffective on small-ticket items.

---

4 Government Rules

The government has the power to make many rules of the game. The government makes tax laws, patent laws, minimum wage laws, superfund cleanup laws, and many, many other laws. These laws govern transactions among all the players in the economy.

In addition to more direct regulation, the government makes the rules that say what rules other players can make. It makes the 'meta-rules' of the game, if you like. This is one role of antitrust laws, which determine, among other things, what types of contracts are legal and what types are not. To our knowledge, none of the contracts we've discussed raise antitrust concerns. But the laws change and interpretations vary, and we recommend that you consult with legal counsel to better understand which contracts to use and which to avoid.

MFCs are a practice that has come under quite intense antitrust scrutiny. In 1979 the Federal Trade Commission (FTC) challenged the use of MFCs by Ethyl Corporation and Du Pont in the sale of lead-based antiknock additives for gasoline. The charge was that the use of MFCs was a so-called facilitating practice, an activity that substantially lessened competition in violation of Section 5 of the Federal Trade Commission Act.[22]

In the first round of proceedings, the commission ruled against Ethyl and Du Pont. It was a controversial decision, with FTC Chairman James C. Miller III dissenting against his own commission's ruling: '[These are] practices that buyers demand . . . The challenged practices . . . arguably reduce buyers' search costs and facilitate their ability to find the best price-value among refiners . . . For these reasons . . . I dissent.'[23]

The case was brought to the New York federal court of appeals, which in 1984 overturned the original FTC decision.[24] The basis of the appeals court ruling was that the use of MFCs had originated at a time when Ethyl was the only manufacturer of the antiknock additives. Since there was no competition at that time, the MFCs obviously had uses other than reducing competition between producers.

We saw these other uses earlier in this chapter. Because MFCs shift the balance of power from the buyer to the seller, they allowed Ethyl to be a tougher bargainer with its customers. That was true when Ethyl had a monopoly, and it remained true once there was competition in antiknock additives. And from the buyers' perspective, MFCs are valuable because they guarantee that buyers won't end up at a cost disadvantage relative to their rivals.

Even if there weren't these other reasons to use an MFC, the FTC's case would have been weak. On the one hand, if your customers have MFCs, you're less likely to go after a rival's customers with low prices. To that extent, competition may be reduced. On the other hand, though, if your customers have MFCs, a rival can more easily come after one of them with a low price. Because you can't match selectively, you may not match at all. To that extent, competition may be increased. The net effect is ambiguous. As we write this book, the status of MFCs continues to be based on the favorable Ethyl–Du Pont ruling.

The use of an MCC is simply part of a contract between buyer and seller, one element of a multifaceted negotiation. While an MFC with one buyer can affect the price another buyer pays, an MCC with a buyer affects the price only that buyer pays. Think of an MCC as a financial option. With a stock option, investors pay dollars today for the right to buy something at a fixed price tomorrow. An MCC is an option that gives the seller the right to sell something tomorrow by matching the best price. This option, just like a stock option, has a direct value, and may be even more valuable if the existence of the option ends up raising tomorrow's price. Of course, anticipating this effect, a buyer who agrees to an MCC can ask for a lower price today in return for providing the option.

One justification for using an MCC comes from the law itself. The Robinson-Patman Act prohibits companies from engaging in price discrimination. According to the law, companies are supposed to sell their products for the same price to all commercial customers.[25] But there's an important exception. An automatic defense against the charge of price discrimination is that you lowered price in response to a competitive bid: '[N]othing herein contained shall prevent a seller rebutting the prima facie case thus made by showing that his lower price . . . was made in good faith to meet an equally lower price of a competitor . . .'[26] An MCC is a way of protecting yourself against the charge of price discrimination.

To our knowledge, there have been no antitrust challenges to take-or-pay contracts or rebate programs. In the case of take-or-pay contracts, the direct economic benefits to suppliers – better planning and prevention of holdup – offer powerful justifications for employing this rule. In the case of rebate programs, the direct economic benefits to the seller – the credit-card issuer foots part of the bill, and joint mailings cut the cost of communicating with customers – again provide a justification for using the rule.

The US antitrust laws can be a bit of a maze. Separate from the way the laws are applied is the logic behind them. The FTC appears to operate from a mental model perhaps more relevant to a bygone era of smokestack industries. It tends to challenge practices that allow companies to sustain prices above variable cost – so-called facilitating practices. That stance doesn't seem to recognize the new economics of the knowledge-based economy.

Pharmaceuticals, software, jet engines, and other knowledge-based products don't fit the traditional economic model. Common to all these businesses are huge up-front R&D costs in relation to the variable cost of making the product. If companies aren't allowed to price above variable cost, they can't earn back their investments. Rules are one strategy that companies can use to help them maintain prices above variable cost and thereby recoup their up-front investments. We're not convinced that current US antitrust law appreciates this perspective.

5 Changing the Rules

Just as the cast of players and the players' added values are important elements of a game, so, too, are rules. Both the significance of rules and the opportunities to change them are often underappreciated.

Too often, people naively assume that 'the rules are the rules.' They treat as if carved in stone rules that actually have little behind them. Herb Cohen, master negotiator and author of *You Can Negotiate Anything*, tells the story of running into a long line at a hotel desk just before the 1:00 P.M. checkout time. Instead of joining the queue, Cohen called up and renegotiated his checkout time to 2:00 P.M. Then he went out and enjoyed a leisurely cup of coffee. When he returned, the line was gone. Cohen's point: don't blindly follow rules.

The freedom to change the rules is a double-edged sword. Don't follow other people's rules blindly; but don't count on others to follow *your* rules blindly, either. Just as you can change the rules or make new ones, so, too, can others.

You can change the rules.

But remember: other people can change the rules, too; don't assume your rules will rule.

That's a good reason not to push a rule too far. For example, if you think that having a meet-the-competition clause gives you a license to keep raising prices to your customers, think again. If a rival comes in with a lower price, the customer is supposed to, even contractually obligated to, give you an opportunity to match. But if your prices were out of line and the rival was able to significantly undercut you, then the customer will quite likely be upset. He'll realize how much he's been overpaying. At that point, the customer may decide not to follow the rules. He may switch without giving you the chance to match. Or he may give you the chance to match and then switch even if you do. What will you do then? You can sue your customer, but that's generally a bad idea, and it certainly won't be good publicity when people discover what brought you to that point in the first place.

Another reason that your rule may be overturned is either that you lose power or that someone else gains power. Before Nintendo arrived on the scene, the rule in the toy business was that retailers placed their orders in January or February, received the product in the summer, and then waited until December to pay their bills. Nintendo changed the rule. It required retailers to place orders, take delivery, and pay in quick succession. Nintendo was able to rewrite the rules because, as we saw in the Added Values chapter, it had the added value.

In 1994 Maurice Saatchi was ousted from the chairmanship of

Saatchi & Saatchi, the advertising agency he had cofounded with his brother. Saatchi's response was to establish his own new agency, M&C Saatchi. He brought with him several of the people at Saatchi & Saatchi who'd worked on the British Airways account. British Airways had to decide whether it wanted to stick with the old Saatchi agency – now called Cordiant – or follow Saatchi to his new firm.

Cordiant found itself facing a new competitor and, worse still, had lost some of its best people to that new competitor. But Cordiant had a rule that gave it some protection, so it seemed. Because the people that Saatchi took with him had had a noncompete clause in their employment contracts, they couldn't do anything that would take the British Airways business away from Cordiant.

So when Maurice Saatchi pitched the British Airways account, he couldn't use any of these people. Instead, he sat life-sized color cutouts of his excluded colleagues around the table. Saatchi then told British Airways that to do the best work possible, he'd need these people on the account. He suggested that British Airways go back to Cordiant and pressure it to drop the noncompete clause.

Cordiant was put in a no-win position. Saying no wouldn't exactly please British Airways. And remember, British Airways was Cordiant's customer – a jumbo-sized one, at that. Turning down a customer's request is hardly a recipe for keeping the customer's business, but saying yes would make the new Saatchi agency a more formidable contender for the account. Either way, the British Airways account would be put at greater risk.

In this game, British Airways had the real power. Saatchi leveraged the power of British Airways and succeeded in changing the rules. He got his people freed to work on the British Airways account. And he got the account, too.

Even when a rule seems firmly established, you always need to remember it might get renegotiated. If you can't control a rule, it's risky to base your strategy on it.

In the marketplace, it's the party with the power who gets to make the rules. A straight flush is almost unbeatable in poker. But, as they said in the Old West: 'A Smith & Wesson beats a straight flush.'

7 Tactics

Perception is reality.
– Bishop Berkeley

Games in business are played in a fog – not von Clausewitz's fog of war, perhaps, but a fog nonetheless. That's why perceptions are a fundamental element of any game.

It's perceptions of the world, regardless of whether they are accurate, that drive behavior. Mike Marn, management consultant at McKinsey, recounts a striking example: 'One price war in industrial electrical products started when an industry trade journal mistakenly inflated the total market volume by 15%. The four major players all thought they had lost market share and dropped prices to recover what was really never lost.'[1]

The job of managing and shaping competitors' perceptions is an essential part of business strategy. In 1994, for example, Rupert Murdoch's *New York Post* cleverly averted a price war with the rival *Daily News* by creating the perception that it was ready to start one. Presently, we'll see how this came about.

Sometimes it's customers or suppliers, not competitors, that need convincing. How can Federal Express absolutely, positively convince people of its reliability? How can a job candidate convince a prospective employer that it won't go wrong by giving him a chance? How can an author convince a publisher that he has a great book to write – and that he'll complete it on time?[2] The need to convince goes both ways. How can the employer convince the job candidate that it will provide valuable training and experience? How can the publisher convince the author that it will invest in marketing the book? These are some of the questions we'll answer in this chapter.

Perceptions play a central role in negotiations. Buyers and sellers often have different views of the pie; sellers portray what they have to offer as valuable, while buyers remain skeptical. Perhaps these are honest assessments, or perhaps they are negotiating ploys. How can buyers and sellers come to an agreement? What should they tell each other? What should they *not* tell each other? Should they try to resolve any difference in perceptions before trying to reach an agreement? We'll also answer these questions and suggest some new ways to negotiate.

The domain of perceptions is universal. Everything is a matter of perception, even perceptions themselves. Reflecting the breadth of this topic, the chapter draws examples from several different spheres of life – the business, the personal, and the everyday.

Change people's perceptions, and you change the game. Shaping perceptions is the domain of tactics. By 'tactics,' we specifically mean actions that players take to shape the perceptions of other players. Some tactics are designed to lift a fog, others to preserve a fog, and yet others to stir up new fog. We'll look at all three possibilities.

1 Lifting the Fog

Some people say it's a jungle out there; or it's a zoo. Either way, we can learn a trick or two from the animals. That's why we start with an example from the animal world before turning to the business world.

The Peacock's Tail One of the puzzles of evolution is why the males of certain species of birds, such as birds of paradise or peacocks, sport such extravagant tails. If it's survival of the fittest, having a long tail would hardly seem to help. At first blush, the tails are a big liability: they are heavy, get caught in bushes, and attract predators. But Charles Darwin had an explanation for these extravagances. Dr Richard Dawkins of Oxford University, zoologist and author of *The Selfish Gene*, puts the Darwinian explanation this way:

> **Females followed a simple rule: look all the males over, and go for the one with the longest tail. Any female who departed from this rule was penalized, *even* if tails had already become so long that they actually encumbered males possessing them. This was because any female who did not produce long-tailed sons had little chance of one of her sons being regarded as attractive. Like a fashion in**

> **women's clothes, or in American car design, the trend**
> **toward longer tails took off and gathered its own**
> **momentum.**[3]

So when it comes to a peacock's tail, beauty isn't in the eye of the beholder. Rather, beauty is what a peahen perceives to be in the eye of other peahens. Attractiveness is a self-fulfilling perception. If the other peahens perceive long tails to be attractive, then any one peahen has no choice, so to speak, but to be drawn to peacocks with long tails. That way, her male offspring will be attractive to the next generation of females. Perception is everything.

The essence of Darwin's explanation of the peacock's tail is that a fashion, once started, feeds on itself. For Darwin, it's accidental that the peacock's tail has turned out to be so extravagant. But zoology professor Amotz Zahavi of Tel Aviv University has an explanation for why long tails are the fashion: they credibly demonstrate superior strength. A peacock with a long tail is telling peahens, in no uncertain terms, that he's a fit male. Dawkins describes Zahavi's view:

> **[Zahavi] suggests that the tails of birds of paradise and**
> **peacocks . . . which have always seemed paradoxical**
> **because they appear to be handicaps to their possessors,**
> **evolve PRECISELY because they are handicaps. A male bird**
> **with a long and cumbersome tail is showing off to females**
> **that he is such a strong he-man that he can survive IN**
> **SPITE OF his tail.**[4]

The tail is a peacock's way of putting its money where its mouth is. A fit male can better afford the risk that a large tail will attract predators. He's also more able to forage for the extra calories it takes to carry a large tail around.

The peacock's tail is a display that distinguishes strong peacocks from strutting pretenders. By proving that they're the ones with the goods, peacocks with long tails succeed in attracting the peahens.

THE FAR SIDE   By GARY LARSON

"Don't encourage him, Sylvia."

Passing the credibility test

The business analogues to the peacock's tail are expensive displays designed to influence perceptions of who you are or what you're likely to do. Rupert Murdoch, publisher of the *New York Post*, knows how to put on a good display better than anyone.

Paper Tiger In the summer of 1994 Murdoch's *New York Post* was test-marketing a cut in the newsstand price to 25 cents and had demonstrated its effectiveness on Staten Island. In response, its major rival, the *Daily News*, raised its price from 40 cents to 50 cents. Under the circumstances, this seemed rather remarkable. As the *New York Times* commented, it was as if the *Daily News* was daring the *Post* to follow through with its price cut across all of metropolitan New York.[5]

There was more going on than the *New York Times* realized. Before it went down to 25 cents, the *Post* had previously raised its price to

50 cents. The *News* had opportunistically stuck at 40 cents. As a result, the *Post* was losing subscribers and, with them, advertising revenue. While the *Post* viewed the situation as unsustainable, the *News* didn't see any problem, or at least appeared not to. A convenient fog. The *News* apparently thought that the *Post* would stick to a 10-cent premium – that it didn't have the will to counter the *News*'s opportunistic behavior.

The *Post* needed to mount a show of strength, to demonstrate to the *News* that it had the financial muscle to launch a retaliatory price war, if necessary. The most credible demonstration would have been to actually start a price war; but that would have been rather self-defeating. The goal was to convince the *News*, without incurring the cost of a full-blown war. So what did the *Post* do?

It put on a display of strength, cutting price to 25 cents on Staten Island. Sales of the *Post* surged, and the *News* discovered that its readers were very willing to switch papers to save 15 cents. It became clear that disastrous consequences would befall the *News* if the *Post* extended its price cut throughout New York City.

It also became clear that the *Post* had the resolve to do just that. The *Post* obviously had the resources to take the small hit caused by cutting price on Staten Island. But the demonstration on Staten Island actually showed more. There's always the risk that a price cut, even a restricted one, will inadvertently trigger a series of tit-for-tat responses leading to all-out war. The *Post*'s demonstration on Staten Island showed that it was willing and able to take this risk. The *Post* was playing a game of brinkmanship. It had the stomach for a fight.

If it had any remaining doubt as to the *Post*'s resolve, the *News* had only to look to London, where just such a meltdown scenario had taken place between Murdoch's *Times* and Conrad Black's *Daily Telegraph*. In September 1993 the *Times* had cut its price from 45 to 30 pence, forcing the *Telegraph* to follow suit. Profits at the *Telegraph* tumbled.[6]

The *Post*'s move on Staten Island wasn't a bluff. The fog in New York lifted, and the *News* saw the light. That's why it raised its price from 40 to 50 cents.

Only the *New York Times* remained in a fog. Murdoch had never wanted to lower his price to 25 cents. He never would have expected the *News* to stay at 40 cents had he initiated an across-the-board cut to 25 cents. The test run on Staten Island was simply a tactic designed to get the *News* to raise its price. With price parity, the *Post* would no longer be losing subscribers, and both papers would be more profitable than if they were priced at 25 cents or even at 40 cents. The *Post* took an initial hit in raising its price to 50 cents, and when the *News* tried to be greedy and not follow suit, Murdoch showed it the light. When the *News* raised its price, it wasn't daring Murdoch at all. It was saving itself – and Murdoch – from a price war.

The lesson of this story is that credibility doesn't come free. You have to put your money where your mouth is. Murdoch spent money to make a point on Staten Island and, given the danger of escalation, put even more money at risk. He showed that he meant business.

The same logic applies when it comes to convincing customers and suppliers that you are what you say you are. The tactic is to put on a display that makes sense only if you're the real McCoy. It's a display that imposters can't, or wouldn't choose to, match. That's why the display means something, and that's why it succeeds in changing perceptions.

A Feather in Your Cap Just as peacocks and peahens engage in mating rituals, so do prospective employers and employees. Only here, the ritual involves résumés, interviews, callbacks, references, and the like. Prospective employers try to gauge the ability of each job candidate. Job candidates try to convince employers of their abilities.

If you're the candidate, educational qualifications are a nice feather in your cap. You've learned a lot in school and college, and that makes you just the person for the job. That's one view. But there's a more cynical view of education: all those qualifications you earned aren't so much a feather in your cap as they are a peacock's tail.

Educational qualifications help employers cut through the fog by helping them judge how smart you are. But it's not that you're smart because you're educated. What you actually learned in college is the least of it. More important is the fact that college wasn't easy. You had to master abstruse academic disciplines. Making it through college is a display of intellectual strength.

This view of education is due to Michael Spence, now dean of

Stanford's Graduate School of Business. So you might expect that Stanford's business school isn't easy. Apparently, it isn't. *Snapshots from Hell* is the title of the book Stanford graduate Peter Robinson wrote describing his first-year MBA experiences.[7]

Stanford isn't inexpensive, either. Annual tuition is more than $20,000, and that's not counting the even greater cost of forgone salary. This high cost of business school sends another convincing signal to prospective employers. Many people say they're committed to a career, but going to business school is a way to prove it. It's an investment worth making only if you're planning to earn it back.

Spending time and money to earn a degree is one way to send a signal to prospective employers. The sort of compensation contract you're willing to accept is another way. In the final stages of an interview for an investment banking job, the candidate is usually asked: how much of your salary do you want to be guaranteed, and how much do you want to be performance-driven? A candidate who elects for a high base salary over a potential bonus sends a negative signal. At best, this person is rather risk-averse; at worst, he lacks confidence in his own abilities. From the investment bank's perspective, that's bad news either way.

The candidate who opts for a low-base/high-bonus package shows that he's willing to bet on himself. He signals that he's confident in his ability to make money for the investment bank — and, in so doing, make money for himself, too. That's what the firm wants to hear.

There's another reason that the bonus-driven compensation package is more attractive to the investment bank. It takes the financial risk out of the hire, regardless of the person's abilities. If the person fails to produce as expected, the firm has to pay the base salary and no more. What if the person ends up earning a huge bonus? That's fine. This happens only when he's made a lot of money for the firm, and then a bonus is a good way to reward and keep a strong performer.

Not shy to try out our own theories, we thought of this analysis of compensation schemes when looking for a publisher for this book. Should we focus on the up-front advance? Or should we try to negotiate a more generous royalty rate? The author's advance is the guaranteed compensation for writing the book, while the royalties are the results-driven bonus. The royalty rate is customarily fixed at 15 percent of sales, with all the negotiation between author and publisher taking place over the size of the advance.[8]

We thought perhaps it would be a good idea to do things a little differently. We could forgo an advance altogether and, instead, ask for a higher-than-usual royalty rate. That would send a signal to publishers that we were very confident of the book's potential. It would also give us a shot at earning more money overall. By taking more of the risk, we would get less if the book failed, but more if it succeeded.

However, there was another consideration. Just as important as communicating our confidence and optimism to the publisher was that the publisher communicate its commitment to us. We needed to know how hard the publisher planned to work at publicizing and marketing the book.

A royalty-only deal takes away a lot of the risk, especially downside risk, for a publisher. So a publisher who agrees to this type of contract says very little about its level of commitment. In contrast, a publisher who agrees to pay a large advance demonstrates confidence in the book. And when it comes to selling a book, confidence can be a self-fulfilling prophecy.

An advance doesn't just convey the publisher's confidence, it changes the publisher's incentives. If the publisher pays a 30 percent royalty, but no advance, its incentive to market the book is diluted by 30 percent. Even the standard 15 percent royalty dilutes the publisher's incentive. But if the publisher pays an advance, it keeps 100 percent of the revenue until the advance is earned out. This way, the publisher has all the right incentives to invest in marketing the book.

There's one more reason why a large advance is good for authors. With a royalty-only deal, the acquiring editor has little personal credibility vested in the project. When the editor has paid a large advance to acquire a book, it's in the editor's interest to make sure that the publisher aggressively promotes the book. That way, the book is more likely to succeed, and the editor is less likely to find his judgment being called into question. A large advance is, in effect, a commitment to the authors that the book won't fail for lack of effort on the publisher's part. That's why we, as authors, decided to focus on the advance, after all.

Similar considerations arise in salary negotiations between job candidates and recruiters. When a candidate agrees to bonus-driven compensation, he may be demonstrating that he's prepared to bet on himself, but the recruiter isn't taking much of a risk in return. If, on the other hand, the recruiter goes out on a limb and hires the candidate at a high base salary, he demonstrates a real commitment. Moreover, with his judgment now on the line, the recruiter has an incentive to see that the new hire works out. Because the new person's success or failure reflects back on the recruiter, the recruiter will help the new hire get experience and opportunities, perhaps even help him get promoted. The recruiter becomes the new hire's guardian angel.[9]

The message of this discussion is that betting on yourself is a display of confidence. So if you really can deliver the goods, bet on it.

Federal Express Should Pony Up Federal Express doesn't make many mistakes. It delivers on its promise of 'The World On Time.' More precisely, it delivers almost every time.

On those rare occasions when FedEx fails to deliver a package as promised, it offers the customer his money back. But when you consider the reason that people send packages via FedEx, that's not much of a guarantee. Often the cost of the postage doesn't come close to compensating someone for the consequences of a late package. FedEx would have a real guarantee if it were prepared to pony up a couple of hundred dollars to the customer whenever it fails to deliver.[10]

Almost all the time, FedEx does deliver packages as promised, so the cost of offering a more generous guarantee would be small. The real cost to FedEx comes from not offering a more generous guarantee. It's a missed opportunity to emphasize the superiority of its service over the post office's Express Mail.

Let's try some suggestive arithmetic: Say Express Mail delivers on time 99 percent of the time, while FedEx has a 99.9 percent success rate. Expressed that way, the difference in performance doesn't sound that big, less than 1 percent. But turn it around and look at failures rather than successes. A 99 percent success rate means a 1 percent failure rate, while a 99.9 percent success rate translates to a mere 0.1 percent failure rate. FedEx is ten times better than the post office at avoiding service failures.

If FedEx were to offer to pay the customer $200 in the event of late delivery, that would cost it, on average, an extra 20 cents per package – the $200 multiplied by 0.1 percent. For the post office to offer the same guarantee would cost ten times as much, or an average of an extra $2 per package.

Right now, FedEx charges quite a bit more than the post office for overnight delivery – $13 versus $10.75. Were FedEx to offer the $200 guarantee, there'd be no way the post office could afford to follow. It would be hard put to absorb the extra $2 cost. But if it tried to pass the extra cost along to the customer, that would erase its price advantage – and that's the only real advantage the post office has to offer.

What's the right size for the guarantee? Should FedEx stop at $200? One concern is that if the guarantee were too large, some people might engage in sabotage, preferring to collect the money rather than their package. It's not inconceivable that FedEx drivers would be put at risk of bodily harm. So FedEx should make the guarantee large enough to get attention, and to represent an honest effort at compensating people for nonperformance, but not so large that some people would actually prefer that the FedEx package not arrive.

If you offer first-class service, you can and should offer a first-class guarantee. It's a tactic that credibly communicates the excellence of your service to customers. Since competitors who offer inferior service will be hard-pressed to match your guarantee, this tactic also helps you stand out from the competition. By the same token, if you fail to offer a guarantee, you miss an opportunity to tell customers how much better your service is.

Offering guarantees has several other benefits.[11] Guarantees are an effective way of committing your organization to provide high-quality service. With the guarantee in place, you'd better deliver or you'll pay the price. Guarantee programs also help alert you to when and where your system breaks down. Instead of taking the time to tell you they're dissatisfied, most customers will brood, bad-mouth you to their friends, or simply walk away. Offering compensation for unsatisfactory service gives customers an incentive to let you know when something goes wrong. That means you get a chance to fix the problem as soon as it happens, and you'll learn how to do better next time. You also get the chance to apologize to the customer – and to show that you mean it by making some recompense. You're most likely to lose a customer when you've just made a mistake and the customer is angry at you. Guarantees let you know when you need to apologize.

A particularly important time to convince people that you're giving them good quality is when you launch a new product. Offering a guarantee is one way to do this. Another way is a low introductory price or free trials, both of which make it cheaper for people to experiment and see whether they like your product. You also signal your confidence that they'll like it enough to come back and buy again at the regular price.

An alternative way to signal your confidence in your product is to spend a lot of money advertising it. The challenge with advertising, of course, is how to cut through the clutter and be credible. You not only have to be heard, you also have to be believed. With the launch of its new Sensor shaving system back in 1990, Gillette found a way to achieve both.[12] The Sensor was a breakthrough product, but Gillette's problem was how to convince people of that fact. Why should they take its claims at face value?

To cut through the fog, Gillette launched a high-profile, high-impact advertising campaign. The ads touted the Sensor's technological advances. But, more important, it was the sort of campaign that got people thinking: 'They're really spending big bucks launching this product. They must be really confident that they've developed a winner. I should try it.' Consumers were right. Gillette spent $100 million on the launch campaign. It would never have done so unless it was confident that people would switch to the Sensor after trying it. Gillette was right, too: people liked the Sensor, and Gillette's worldwide razor sales jumped by 70 percent.

When banks, consultancies, and other professional firms set up operations in new cities, they often spend a lot of money on conspicuously lavish office space. Likewise, MBA students often wear expensive attire to job interviews. In each case, spending big is a way to signal confidence. You don't do it unless you really think people will find your services valuable and will hire you. Whether it's paying for a lavish advertising campaign, lavish office space, or a lavish wardrobe, 'burning' money in a highly visible fashion demonstrates your faith in your product.

Failing the credibility test

Intentional or not, everything you do sends a signal to others. For the same reason, everything you don't do sends a signal, too.

In the story 'Silver Blaze,' Sherlock Holmes is called in to investigate the mysterious disappearance of the Wessex Cup favorite just a few days before the big race. Evidently someone has crept into the stables and abducted the horse. But who? And how did he elude the dog guarding the stables?

> Inspector Gregory: Is there any other point to which you would wish to draw my attention?
> Sherlock Holmes: To the curious incident of the dog in the night-time.
> Inspector Gregory: The dog did nothing in the night-time.
> Sherlock Holmes: That was the curious incident.[13]

Holmes deduces that the villain must have been no stranger to the dog. In fact, the villain was the horse's trainer.

Like Holmes, you can learn just as much from the things that don't happen as from those that do. You have to learn to listen for what it is that you haven't heard.

The Dog That Didn't Bark A large manufacturing company was trying to locate a toxic waste recycling plant. It had picked a certain midwestern town for the site and presented its case to the local inhabitants. The company made three promises: to bring much-needed jobs to the community; to invest in improving the local schools; and to build a plant that would be completely safe.

Still, the locals were unconvinced. What if there was a health hazard, after all? That would be yet another blight on what was an already depressed town. Property values would fall even further, and if people wanted to leave, they'd be unable to. The mere perception that there could be a health hazard was a problem. In order to be in front of the queue, people might sell their houses right away, and that could start a downward spiral in property values. The risk of falling property values tomorrow could precipitate a fall today.

There was a solution. If the company's word was good, there would be little danger of property values tumbling. In fact, the contrary would be true. With more employment opportunities and better schools, property values would almost surely rise. So the answer for the company was to indemnify residents against a fall in their property values. It could have hired independent appraisers to assess real-estate values, not taking into account the proposed plant. Then the company could have announced that it stood willing to buy anyone's house in five years' time at the current appraised value.

Five years would be long enough for any uncertainty over the safety issue to be resolved. In the meantime, the promised benefits from increased employment and investment in the community would become apparent. People would see their property values rise, not fall. No one would want to take up the company's offer to buy them out at the previously appraised value, and the guarantee would end up costing the company nothing.

When this tactic was proposed to the company, management declined to go along. Their behind-the-scenes response was: 'We couldn't do that. It would cost us a fortune. Everyone would take advantage of our offer.' At this point, it became quite clear why the company had a credibility problem. Its refusal to offer a guarantee spoke volumes. How can you convince others if you're unwilling to bet on yourself?

Here's a summary of how you establish credibility for yourself and recognize it in others.

THE CREDIBILITY TEST

1. If you have the goods, put your money where your mouth is:
   - Accept a pay-for-performance contract.
   - Offer a guarantee.
   - Give free trials.
   - Advertise.
2. What you don't do sends a signal, too.
3. Ask other people to demonstrate their credibility to you:
   - Propose a pay-for-performance contract.
   - Ask for a guarantee.
   - Request a free trial.

---

2 Preserving the Fog

Once you've managed to convince people that you have the goods, now what? The next challenge is how to maintain this perception. You now have more to lose than gain if people should get new information that leads them to revise their view of your abilities. You have an incentive to prevent new information from coming out.

Hiding information

We've already seen a little bit of why people might want to preserve a fog. An editor doesn't want his judgment to be in doubt. If he can persuade the publisher to spend enough money promoting a book, the book isn't given a chance to fail and the editor can't be faulted for having acquired it in the first place. No one learns whether the book would have succeeded or failed in the absence of a big push. Recruiters have similar incentives to make their selections succeed and not have their judgments called into question.

Anytime you say yes to a project, you'll be judged according to how well things turn out. Likewise, when you say no, you'll also be judged, but only if someone else says yes.

ET – the Wrong Call In Hollywood, movie executives are known by their successes. For example, Jeff Katzenberg is famous for having developed *Beauty and the Beast*, *Aladdin*, *The Lion King*, and other Disney hits. Robert Newmyer and Jeff Silver built a reputation for producing surprise hits such as *sex, lies, and videotape* and *The Santa Clause*.

Harder to gauge is how many good projects a movie executive has turned down. What were the missed opportunities? For the most part, Hollywood executives are masters at keeping this information hidden. They have nothing to gain and everything to lose by letting it out.

Hollywood mogul Frank Price has had many big successes. But he'll be forever known as the person who gave away ET for $100,000. As the then-president of Columbia Pictures, Price owned the rights to both ET and *Starman*. His view was that there wasn't space for two movies about extraterrestrials in one year. Price was right. But he made *Starman* and sold ET to Universal. *Starman* went on to gross $29 million, ET topped $400 million.[14]

The failure of *Starman* wasn't the problem. No one bats a thousand. The problem was the stunning missed opportunity of ET. Had ET never materialized, no one would have been the wiser. People already thought Price had good judgment, so he had nothing more to prove. But the astronomical success of ET proved that Price had made the wrong call.

Whenever you turn down a project, you have a reason to hope it never sees the light of day. Granted, if someone else picks up the project and fails, your decision is vindicated. But you seldom earn brownie points this way. When you turned down the project, you were probably given the benefit of the doubt, anyway. On the other hand, if someone picks up the project and makes a success of it, your judgment can now be faulted. Once you decide not to gamble on something, you have little to gain and much to lose from having people learn whether the bet would have paid off. You're better off if the fog is preserved.

Following the Herd The fear of being proved wrong leads people to move in herds. The scene is a familiar one. First, all the pension funds buy IBM; then they all sell IBM.[15] First, there's a merger wave as every company scrambles to become a conglomerate. Then the wave crashes and everyone divests unrelated businesses because 'focus' is the new wave.

Economic forecasters all seem to come up with remarkably similar predictions – and, most often, all remarkably wrong. One reason is that they all use similar models and the same historical data. Another reason is that forecasters fear going out on a limb and being proved wrong. London Business School professor and leading business strategist John Kay explains the phenomenon:

> **Even when they are proved wrong, forecasters see it as important to maintain the consensus in retrospect. For example, banks maintain as an article of faith that the depth of the recent [UK] recession and the magnitude of the property market collapse could not have been predicted. If it could have been, those responsible for the lending excesses of the 1980s would be guilty of gross negligence rather than helpless victims of events ... [I]t is often more important to be wrong for the right reasons than to be correct.[16]**

If you follow the herd, you'll succeed or fail along with the herd. The fog is preserved. You'll never stand out if you're right, but you're also less likely to get eaten alive if you're wrong. If it turns out that you've made a bad decision, you can say: 'Who could have known? Look at all the other people who made the same decision. At the time, everyone thought it was the best thing to do.' With luck, no one will question your judgment.

The desire to preserve the fog can cause otherwise-hard-to-explain behavior. As teachers, we've noticed that some students don't study for tests precisely when they're in the most danger of failing. Obviously, they'll be much more likely to fail, but now they can explain the failure away as the result of not studying. They avoid having to do much soul-searching. People set themselves impossible tasks for exactly the same reason: when they fail, they don't have to question their abilities. Everyone, themselves included, remains in a fog.

Even if there's a 90 percent chance that lifting the fog will reveal something good, the 10 percent chance that something really bad will come to light may be too much of a risk to take. That was the position taken by Continental Corporation, a financially troubled insurance company.

CNA Insurance had just made a bid for Continental, substantially outbidding the current contender, Insurance Partners. As a condition of its bid, CNA asked to engage in extensive due diligence. If the audit didn't reveal any major problems, CNA would proceed with the deal. Otherwise, it would withdraw its offer altogether. In spite of the attractive price, Continental's board of directors rejected CNA's bid:

On November 17, 1994, the Board of Directors met and reviewed in detail the CNA proposal ... and various risks inherent in CNA's request to conduct intensive due diligence, including the potential adverse effects a possible decision by CNA (following such due diligence) not to make an offer could have on market and rating agencies' views of the company and on the willingness of Insurance Partners to proceed with its transactions. The Board of Directors did not accept the CNA proposal because of the Board's concerns with respect to such potential adverse effects. (Proxy Statement to Shareholders, the Continental Corporation, March 29, 1995)

The board wasn't prepared to take even the small risk that CNA, having done its audit, would then say no. It feared everyone would infer that CNA had discovered some very bad news. Insurance Partners, and any potential bidders in the wings, would be scared away. Rating agencies might downgrade Continental. Customers might lose confidence and take their business elsewhere. A disaster.

In turning down CNA's bid, Continental's board increased the chance the deal would fall through. It did so in order to preserve an essential fog. In the end, CNA dropped its demand to do extensive due diligence, and the deal went through. Continental was lucky. CNA was also lucky, since no problems were uncovered.

HIDING INFORMATION

Having made a favorable impression, people try to preserve it by:
- burying projects they've turned down
- following the herd
- creating reasons to fail

Negotiation tactics

The typical negotiation takes place in a fog. People tend to get lost navigating through this fog, and the negotiations run aground. One mistake is to overstate what you need, hang tough, and thereby kill the deal. Another mistake is trying to strengthen your hand by revealing something that, in fact, would have been better left hidden. A third mistake is forcing a consensus when preserving differences of opinion would actually be more helpful in crafting a deal. In this section, we'll look at these three negotiation traps and suggest some ways out.

As the Escrow Flies Negotiations can be full of bluffing and posturing. People make extreme demands to try to anchor the subsequent back-and-forth in their favor. In this kind of climate, if you reveal what you really need, you can find that information being used against you. There's a clear incentive to act tough. But if everyone acts tough, it gets much harder to craft agreements. Negotiations deadlock, and then the deal doesn't get done, even though everyone would benefit if it did.

Let's look at this problem a little more closely. You've got something to sell, and you're talking to a potential buyer. Your absolute floor is $100. Any less and it makes more sense for you to go elsewhere. In an attempt to be reasonable and fair, you ask for $120. The buyer turns you down. Even worse, now he knows that your floor is below $120. All of his counteroffers will be below $120, and you may have to settle for an amount very close to your floor of $100. That doesn't seem reasonable or fair.

The next time you're in this situation, you decide to ask for a lot more than $100. You demand $180, and the buyer comes back with an offer of $140. You stand firm, hoping to get more. In fact, the buyer's ceiling is $150, and his $140 offer was made in good faith. He simply can't pay your price. By refusing to budge, you've killed the deal altogether.

Asking for a little gets you a little, and holding out for a lot may get you nothing. Neither way is clearly right.

The problem is not with the players, but with the game. University of Chicago Business School professor Rob Gertner and NYU Law School professor Geoffrey Miller have come up with a better game to play. They've devised some ingenious rules that enable people to behave reasonably without having their lunch eaten. They call their negotiation method 'settlement escrows.'[17]

Here's how settlement escrows work. The buyer and seller agree to bring in a neutral third player to act as mediator. The seller tells the mediator, in private, a price at which he'd be willing to sell. Likewise, the buyer lets the mediator know, again in private, a price at which he'd be willing to buy. The mediator checks to see whether the two prices cross – that is, whether the buyer's offer exceeds the seller's bid. If so, the mediator calculates the midpoint price, and seller and buyer transact at that price. If the two prices don't cross, the mediator doesn't reveal either price. He announces only that the prices didn't cross.[18] Neither side learns the other's bid, and the two parties can go on negotiating without prejudice.

Let's return to our example. It's now much safer for you, the seller, to ask for $120. If the buyer quotes the mediator a price above $120, the deal is done at the midway price. Thus, if the buyer quotes $160, he ends up paying you $140. That's fine by you. You get more than you asked for. And while the buyer now knows you asked for $120, and may be kicking himself for not quoting the mediator a lower price, it's too late for him to do anything about it. The game is over. Those are the rules. As the seller, you're protected.

What if the buyer quotes the mediator a price below $120, say $110? Then the deal doesn't go through. True, you and the buyer will have to try some other way to reach an agreement. But in making a reasonable opening demand, you haven't compromised your position in any subsequent negotiations. All the mediator reveals is that the prices didn't cross. The buyer knows that you asked for a price above $110, but that's all he knows. He doesn't know whether you asked for a little more or a lot more. Since the buyer doesn't have the information he'd need to box you in, you're protected once again.

Settlement escrows allow people to negotiate from behind a veil. Ordinarily, when you make a demand, you reveal your hand. Settlement escrows preserve the fog. You can say what you really need without giving away much information. When the parties in a negotiation feel safe enough to make reasonable demands, they're much more likely to reach an agreement. There's a much better chance that whenever there's a mutually beneficial deal to be made, it will be made.

Gertner and Miller conceived of settlement escrows as an aid to pretrial negotiations. You may be willing to pay $100,000 to settle the matter, but you may not want the other side to know that – unless it's willing to settle here and now. If it isn't, revealing the fact that you're willing to settle for $100,000 may be what tips the other side into deciding to go to court rather than to continue negotiations. The solution is for both parties to agree, at the outset, to use a settlement escrow.

The value of settlement escrows obviously isn't limited to the legal arena. The idea could fly in a wide range of situations. You're willing to pay a lot for an employee's services, a parcel of land, or a patent, but you don't want to tip your hand. The employee is willing to work for very little, the landholder is anxious to sell, or the inventor is keen to see his idea commercialized, but none of these people want to tip their hands, either. In all these cases, use of a settlement escrow would maintain a veil over the negotiations, allowing both parties to negotiate in good faith.

In a negotiation, what you know and what other people know aren't the only things that matter. Do you know what others know? Do others know what you know? Do others know how much you know about what they know? The fact that you know something that others know, too, has very different consequences when others know that you know it.

The Cat in the Bag I have something unpleasant on my mind. I suspect that you know what's on my mind. But do you know that

I suspect you know? I believe that you might, but I don't know for sure. There's room for doubts, and that may be the best way to leave things. Some thoughts are better left unspoken.

Everything changes when I reveal what's on my mind. Now you know for sure what I'm thinking. And you know that I know you know. You even know that I know all this. The fog lifts and the truth can no longer be hidden. As they say, the cat has been let out of the bag. And that's a problem, because there's no easy way to get a cat back in a bag.

In a marital dispute, the cat in the bag may be the threat of divorce. You may have it in the back of your mind to ask for a divorce if things can't be worked out. You may believe that your spouse suspects as much. You may even think that your spouse senses that you believe your spouse suspects as much. But it's still better to leave the threat unspoken. However serious the problems you and your spouse are having, there's always a chance of working things out, provided the two of you are committed to trying. But once the threat of divorce is made explicit, it becomes much harder to make that commitment. Any doubts about what's on your mind are now gone. You've just revealed that you're contemplating life beyond the marriage. How can your spouse remain committed to trying to make the relationship work, knowing you've already got one foot out the door?

If you threaten a divorce unless certain conditions are met, you risk getting the response: 'Oh, you want a divorce, do you? You got one.' You want the marriage, albeit with some changes, not a divorce. But it's too late. The threat of divorce, now made explicit, can be self-fulfilling. It is better left implicit.

We began this book by asking: 'If business isn't war and it isn't peace, what is it?' We said it's war and peace. But one participant in a workshop we ran answered: 'It's marriage.' He had a point. There are elements of competition and cooperation in every relationship, business or personal. And our experiences in one domain help us understand the other better. In business relationships, just as in marital relationships, some thoughts are better left unspoken.

Business negotiations involve both promises and threats. Some of those threats, however, are better left implicit. We witnessed a case where a supplier, frustrated by the slow pace of negotiations over a contract renewal, threatened to cut off the buyer if he didn't agree to the supplier's terms. The buyer gave in to the supplier in the short term. He had no choice. But the damage was done. The buyer, seeing that the supplier didn't have a problem inflicting harm on him, could hardly continue the relationship. The buyer set out to ensure that he would never again be caught in a vulnerable position. He found a new source of supply, and even a backup to that new source. As soon as he could, the buyer stopped doing business with the original supplier.

What should the supplier have done? He should have allowed the buyer to come to his own realization about what might happen if negotiations deadlocked. If the buyer continued to drag his feet, the supplier could have suggested bringing in a mediator. A skillful mediator knows that part of his job is to help each party see the consequences of failing to reach agreement. A good mediator would have helped the buyer see that if he pushed too hard, the supplier might well cut him off. The buyer needed to face this fact, but when the supplier made the threat to cut him off, he let the cat out of the bag, and that was the beginning of the end of the relationship. It would have been much better to have a third party help the buyer peek in the bag and see that there was a cat inside. That might well have been all that was needed to stop the buyer from dragging his feet.

Supplier and buyer needed to maintain a fog over what could happen if the negotiations broke down – not necessarily a thick fog, but definitely some fog. The use of a mediator would have helped preserve a mutually convenient fog.

When inexperienced negotiators get frustrated by the slow pace of discussions, they often start making more explicit threats. That's a mistake. If you're not sure whether you can hold your tongue, think about bringing in a mediator.

Negotiations are about coming to an agreement, but that doesn't necessarily mean that everyone has to see things the same way. Agreements can be reached even when people stick to their differing perceptions. Indeed, differences of opinion can actually make it easier to reach an agreement. The following disguised story of a fee negotiation between an investment bank and its client shows how this works.

Disagreeing to Agree The client was a company whose owners were looking to sell. The investment bank had identified a potential acquirer. So far the investment bank had been working on good faith; now it was time to sign a fee letter.

The investment bank suggested a 1 percent fee. The client figured that its company would fetch $500 million and argued that a $5-million fee would be excessive. It proposed a 0.625 percent fee. The investment bankers thought that the price would be closer to $250 million and that accepting the client's proposal would cut their expected fee from $2.5 million to about $1.5 million. Ultimately, one side would be proved more right than the other as to the market value of the company. But, right now, there was a fog.

Naturally, the investment bank thought it knew best. It could have tried to convince the client that a $500-million valuation was unrealistic and that its fear of a $5-million fee was therefore unfounded. The problem with this approach, though, was that the client didn't want to hear a low valuation. Faced with such a prospect, it might have walked away from the deal and even from the bank altogether – and then there would have been no fee.

The client's optimism and the investment bankers' pessimism created an opportunity for agreement rather than argument. In the end, both parties agreed to a 0.625 percent fee with a minimum guarantee of $2.5 million. That way, the client got the percentage it wanted and considered the guarantee a throwaway. With a 0.625 percent fee, the guarantee kicked in only if the sale price was below $400 million, and the client expected the price to be $100 million above that. Because the investment bankers had expected $2.5 million under their original proposal, now that this fee was guaranteed, they could agree to the lower percentage.

Negotiating over pure percentage fees is inherently win-lose. If the fee falls from 1 percent to 0.625 percent, the client wins and the investment bankers lose. Going from 1 percent to 0.625 percent plus a floor was win-win – but only because the two parties maintained different perceptions.

The negotiations between the company for sale and its advisers were just the warm-up to the real negotiations between the company and its ultimate buyer. Here again, differing perceptions proved to be mutually beneficial.

The company's owners thought that the business was likely to continue growing at 10 percent per year, thus justifying their $500-million asking price. The buyer forecast flat growth and offered $250 million.[19] The buyer might have tried to convince the owners that their rosy scenario was all wet, but that would have been a mistake. Instead, the buyer used the difference in perceptions to help forge an agreement.

The buyer offered a mixture of cash now and delayed payments based on the company's future performance. If the company's growth was flat, the total price would be low. If the company continued to grow as it had, then the seller would get what he was asking for. The benefit of preserving the fog was that each side had a different view of the agreement: the buyer thought he was paying a little, while the owners thought they were getting a lot.[20]

NEGOTIATING IN A FOG

MISTAKES

1. Revealing the minimum you need. You risk getting exactly that, and no more. Posturing is no solution; that risks deadlock.
2. Making threats explicit. Even if a threat is already implicit, making it explicit changes perceptions. There's no going back.
3. Trying to resolve differences of opinion between you and the other party. This is hard to do, and possibly counterproductive.

SOLUTIONS

1. Establish a settlement escrow to promote good-faith negotiating by both parties.
2. Bring in a mediator to help the other party understand the consequences of nonagreement.
3. Recognize what you and the other party do – and don't – have to agree on. Use differences of opinion to structure win-win deals.

---

3 Stirring Up the Fog

If you can't convince 'em, confuse 'em.
– Harry Truman

Simplicity is a virtue – sometimes. Other times you need to make things complicated, even unpredictable. You need to create a fog. A simple game quickly becomes transparent, and you may not always want people to see through what you're doing.

In poker, you're unlikely to win a big pot if you bet only when you have a strong hand. That's because, after a while, the other players will see through what you're doing. There'll be no fog. They'll realize that whenever you raise, it's because you have a strong hand, and so they'll fold. That gives you an opportunity to bluff and win more small pots. You can raise with a weak hand and trick the other players into folding, but you don't want to get caught bluffing. Or do you? Professor Tom Schelling, a leading game theorist, has pointed out that there can be a bigger gain from bluffing and getting caught. Now you've really stirred up the fog. If people catch you raising on a weak hand, they'll be much more willing to challenge you in the future when, unbeknownst to them, you raise on a strong hand.

If people understand what you're up to, the results can be self-defeating. Thus, the Internal Revenue Service has to keep its audit formula a secret. If people knew just what would trigger an audit, then dishonest taxpayers would be better able to escape detection. The fact that the IRS's audit formula is kept hidden makes it a more effective enforcement mechanism. The same logic explains why companies carry out random drug testing of employees or perform surprise internal audits. Unpredictability is the key to effectiveness.

On Wall Street, some people make their livelihood from creating complexity. They take a simple financial instrument, such as a mortgage, and split it up into half a dozen or so esoteric components that can trade separately. By turning one market into half a dozen, the market makers greatly increase the scope for trading. Because each component trades independently, it's easy for prices to get out of line. If the components are collectively underpriced, traders can buy up all the parts, put them back together to re-create the original instrument, and sell that at a profit.[21] Yale School of Management finance professor Steve Ross has a pithy way of explaining why traders value complexity: 'Not everyone realizes that pulp, seeds, peel, and juice can be combined to form an orange.'

Creating complexity

Complex pricing schemes create a fog that obscures the true price. From the seller's perspective, that's sometimes just as well. The Nintendo Entertainment System premiered at a price of $100. That sounds cheap enough. But parents soon found themselves in for a penny, in for a pound. Buying the machine was only the beginning. After that came the game cartridges. On average, families bought eight or nine cartridges for the machine, at $50–$60 each. The life-time cost? Around $550. Had parents been more aware of this number at the outset, perhaps they might have better resisted their kids' pleadings for a Nintendo.

Microsoft Windows 95 debuted at $85. A true bargain? Yes, but not quite as good as it looks. People soon discovered that there were several hidden costs. To run Windows 95 effectively, they had to buy more memory ($360), a bigger hard disk ($200), and a faster microprocessor ($300). The cost of the software was the proverbial tip of the iceberg: 90 percent of the costs were hidden.

People get particularly upset if sellers hike prices in response to a spike in demand. That's why sellers sometimes try to hide price rises in a fog. In college towns, there's always a shortage of hotel rooms on graduation weekend. Instead of raising price, some hotels sell only four-day packages. Parents coming into town to see their son or daughter graduate might want a room for one night, but they have to buy three extra nights. The net effect is no different from a 300 percent increase in room rates, but people apparently don't perceive it this way, perhaps because they get the option of staying the extra nights. In Geneva, participants in the Telecom '95 conference saw this practice taken to a new extreme. Responding to a flood of requests for accommodations, an ideally located estab- lishment – we'll call it Hotel Noah – offered rooms at the apparently reasonable rate of $350 per night. Minimum stay: forty nights. Unbelievable! One skillful negotiator reported: 'I managed to negotiate the terms down to a minimum of twenty-six nights.'[22] Not bad, but the conference lasted only ten days.

You might imagine that hotels don't create much fog with these tactics. After all, they're not much more than thinly disguised price increases. Still, creating even a little fog can be much better than acting in a way that's plain for all to see. Following the destruction caused by Hurricane Andrew in 1992, Georgia-Pacific jacked up its lumber prices in Florida. Georgia-Pacific appeared to be profiting from others' misfortune. As a result, the company found itself at the center of a storm of its own making. There was immense public criticism, even an investigation by the Florida Attorney General's Office. Georgia-Pacific would have done better had it raised price only a small amount and, as a quid pro quo, gotten long-term contracts from its customers – effectively saying: 'I'll sell to you in times of shortage if you'll agree to buy from me in times of plenty.' This would have been a less naked exercise of power.

Most of the time, complex pricing schemes serve to mask high prices. Sometimes, though, sellers make things complicated in order to hide how little they're charging rather than how much. They're willing to sell their product at a bargain price, but they hide the low price to avoid creating a perception of low quality.

When it first came out, Microsoft's Powerpoint presentation software was a distant second to Harvard Graphics. Even after Microsoft improved its product to the point where it outperformed the competition, people were slow to switch over. To build sales, Microsoft could have tried cutting the price of Powerpoint, but people might not have believed that a discounted program was better than the $290 Harvard Graphics. So Microsoft kept the list price of Powerpoint high, at $339, but also included the program along with the much more popular Word and Excel in its Office software suite. Thinking that they were getting a $339 program as a bonus, people were eager to try it. Today Powerpoint is the clear leader in presentation software.

Complex pricing schemes make it hard for buyers to comparison shop. That's a problem not only for buyers but also for entrants trying to break into the market. If buyers can't quite figure out how all the prices add up, how can they decide whether it makes sense to switch? They can't, so, for the most part, they don't.

Making the Right Call What do you pay for a minute of long-distance calls? Not sure? We thought so. Most people find it impossible to keep track of all the different rates they're charged. There are the rates for daytime calls, evening calls, in-state long-distance calls, out-of-state long-distance calls, international calls, operator-assisted calls, and more. It's a big fog.

In Japan, there's another layer of complexity. The phone companies there even charge for local calls. This pricing fog is a problem for DDI (Dai Ni Den Den Inc., or '2nd Telegraph and Telephone') as it tries to gain market share against the dominant carrier, NTT (Nippon Telegraph and Telephone). For some calls, DDI is cheaper; for others, NTT is cheaper. But which carrier is cheaper overall? That all depends on your calling pattern, and that's hard to predict. There isn't a compelling case to switch to DDI. If you know that a particular call would be cheaper on DDI, you can route the call that way by using a four-digit prefix, but that's a bother.

DDI has found a way to solve this problem. In partnership with Kyocera, a major shareholder, DDI has developed a chip that goes inside the phone. The chip stores the prices that DDI and NTT are charging and automatically routes each phone call via the lower-cost provider.[23] Whenever there's a price change, DDI simply sends the new information over the lines to the chip inside the phone. Now consumers don't have to worry about the fog at all – the chip does the comparison shopping for them. And DDI is sure to get their business whenever it has the better price.

The 'chip phones' are the essential complement to DDI's business. And just as the chip phones increase the value delivered by DDI's service, DDI's low rates make these phones more valuable. This is why electronics manufacturers such as Matsushita and Sanyo are happy to build the chip phones. They can make a selling point of the fact that these phones more than pay for themselves with the money they save on phone bills. So Matsushita and Sanyo help DDI get the chip phones inside people's homes.

Complex pricing schemes have a number of costs. Customers can get confused and frustrated trying to peer through the fog, and that damages their perception of the product. In the case of airline pricing, the development of yield management systems in the 1980s led to a huge number of different fares in the marketplace. By the early 1990s the complexity was out of control: American Airlines alone had nearly half a million fares. It was an administrative nightmare. The computers may have been able to keep track, but no human – whether traveler or travel agent – could.

High-Altitude Fog In April 1992 American Airlines tried to dispel the fog with its Value Pricing initiative. Bob Crandall, CEO of American, explained the move:

> **In our unsuccessful pursuit of profits, we have made our pricing so complex that our customers neither understand it nor think it is fair ... By moving to a new approach, which emphasizes simplicity and equity and value, we hope to regain the good will of our customers ... We call it Value Pricing.**[24]

Value Pricing involved a dramatic simplification of the fare structure. From now on, there would be just four kinds of fares: first class, regular coach, and two discount fares. The other airlines responded positively. Within forty-eight hours of American's move, United put together its own Fair Fares price-simplification scheme. Alaska, America West, Continental, Delta, Northwest, and USAir were also quick to follow American's lead and adopt simplified pricing. An Alaska Airlines representative exclaimed: 'If we were a big airline, it's something we would have done ourselves. It's dynamite.'[25]

Imitation of Value Pricing was healthy. The more airlines that copied American, the simpler airline pricing got. That meant fewer disgruntled travelers and travel agents, and that, of course, was good news for the airlines. Value Pricing changed the game in another way. When the pricing game was being played in a fog, there was always the temptation for an airline to engage in furtive price cutting. An airline could hope to cut price and steal some share before the other carriers had time to spot what was really happening and respond. With Value Pricing, the game was much more transparent. The chances of pulling off a surreptitious price cut were reduced, and thus the incentive to try was reduced, too. Simplified prices meant more stable prices – a clear benefit to the airlines.

TWA was the spoiler. It saw Value Pricing as an opportunity to cut price and steal share. Apparently, it reckoned that American, with only four fares to play with now, would be less likely to respond. So just three days after American's announcement, TWA came out with fares 10–20 percent below American's. In response to TWA's move, America West, Continental, and USAir matched TWA's price cuts, and a week later, American felt obliged to follow suit by cutting prices across the board.

Over the following months, Value Pricing lost momentum as more airlines started coming up with discounts and special promotions that nibbled away at the edges of simplified pricing. For example, Northwest came out with a promotion to encourage family travel. Its 'Grown Ups Fly Free' deal offered a free ticket to adults accompanying children.

By September 1992 American admitted that Value Pricing had stalled, and decided to return to the previous status quo. Why did Value Pricing fail? The financially distressed airlines – principally TWA and Northwest – had pressing short-term cash needs and couldn't resist the temptation to cut price. They couldn't afford to wait around for the long-term benefits that Value Pricing promised. Another reason was that during his tenure at American, Bob Crandall hadn't exactly won industry popularity contests. No matter how good an idea Value Pricing was, some wouldn't go along with it simply because it was Crandall's baby. This was the lose-lose mindset that helps explain how the US airlines managed to lose nearly $5 billion that year.

COMPLEX PRICING SCHEMES

1. Hide high prices.
2. Disguise opportunistic pricing.
3. Hide low prices, too, preserving an image of quality.
4. Hamper comparison shopping

. . . but also . . .

1. Increase administrative costs.
2. Confuse and frustrate customers.
3. Encourage furtive price cutting by competitors.

Shaping opinions

Many games are ultimately decided in the court of public opinion. Here, perceptions aren't just part of the game, they're the whole game. In this section, we'll look at how to play this game and how not to.

CBS Gets a Black Eye In the last weeks of the Bush presidency, Congress overrode the president's veto – for the first and only time during his administration – to pass the Cable Television Consumer Protection Act of 1992. Cable television had been partially deregulated in 1984, and since then rates had been rising at three times the inflation rate. The 1992 Act was designed to reregulate cable. It also contained a little-noticed provision allowing the nation's 1,152 commercial broadcast stations to choose between 'must-carry' status and 'retransmission consent.'

Stations choosing must-carry status were guaranteed a slot on cable but then couldn't charge cable operators for carrying their programming. Stations choosing retransmission consent gave up an automatic right to be carried but were free to negotiate a fee for their programming. If no deal could be struck, the station could then withhold its consent, and the local cable operator would have to stop retransmitting the station's signal. Most of the ABC, CBS, NBC, and Fox stations chose the retransmission-consent option.

Broadcasters had lobbied for the retransmission-consent provision to be put into the 1992 Act. And when the bill was passed, they were very enthusiastic. CBS Chairman Larry Tisch boasted that television stations could potentially 'wring $1 billion a year in royalties from cable.'[26]

Cable operators, led by Tele-Communications, Inc. (TCI), and Time Warner Cable, declared that under no circumstances would they pay for the broadcasters' signals. Nor were they willing to pass these costs along to their customers.[27] 'We will listen to any scheme that adds value to the consumer. We are not willing to do a zero-sum game of transferring wealth from our consumers to CBS Chairman Larry Tisch,' was the response of TCI Chairman John Malone.[28]

To present its case to the public, Time Warner Cable hired communication strategists Shepardson Stern and Kaminsky. Lenny Stern explained to us how he stirred up some fog. His market research showed that consumers had a visceral reaction when told of Tisch's billion-dollar boast. They fully expected that Tisch's billion would come straight out of their pockets. When asked to estimate the damage, people feared they'd have to pay an extra $15 a month in cable fees. It was all seen as extremely unfair: why should cable subscribers have to pay for something their neighbors with rabbit ears could get for free off the air?

Whether customers were doing their arithmetic right or wrong, or whether paying the networks was really fair or unfair, was beside the point. As always, the game was all about perceptions. By tossing out the billion-dollar figure without explanation, Tisch had left the door wide open for people to put their own spin on the situation. Time Warner Cable and Stern seized the opportunity to cast the conflict in a light that favored the cable companies: if broadcasters got their way, consumers would pay. On the opposite page is a representative ad from their campaign.

Tisch would have done better had he framed things differently. Consumers' fears of an extra $15 a month in cable fees were ungrounded. There are 60 million cable households. On a per-subscriber basis, Tisch's billion translated into an extra payment by the cable companies of just 35 cents per month for each of the four networks. That's the number Tisch should have emphasized. As for fairness, Tisch could have pointed to cable-only networks such as CNN and ESPN, which were charging the cable operators monthly

SOON YOUR FAVORITE SITCOM COULD COST YOU SO MUCH IT ISN'T FUNNY.

TELL THE NETWORKS YOU'RE NOT GOING TO PAY FOR FREE TV.

CAST YOUR VOTE FOR FREE TV. CALL 1-800-FREE TV 3.

fees of 10–40 cents per subscriber, depending on popularity. CBS, which was much more popular than any of the cable-only networks, would be a bargain at 35 cents a month.

Tisch might even have turned the tables. Up to that point, the four networks were giving away their programming to the cable companies, who were then charging subscribers up to 75 cents a month per channel. Tisch could have publicly challenged the cable

225

companies: 'I'll continue to provide CBS programming for free if you lower the customers' bill by 75 cents a month.' If the cable companies had rejected this challenge – and we suspect they would have – then Tisch could have argued that it was only fair for CBS to get a share of the 75 cents.

The networks did try to make some of these arguments, but it was too little too late. The cable companies had already succeeded in painting the networks as the bad guys. When the cable companies and the networks came to negotiate fees, the networks were in a weak position. They'd be the ones the public would blame if the negotiations broke down.

In the end, all four networks gave retransmission consent for free, although three of them did manage to get a consolation prize. Fox played its hand early and did the best. In June 1993 TCI and Fox announced their deal: Fox gave retransmission consent, while TCI agreed to pay Fox 25 cents per subscriber for a new, undefined cable channel. ABC and then NBC got similar, though less generous, deals from the cable companies. Each was paid for creating a new cable channel.[29] CBS held out the longest – and walked away with absolutely nothing, except perhaps a black eye:

Having rejected cash, and now a CBS cable channel, we are at a loss as to what the cable industry does want, short of our abject surrender.
– Larry Tisch, Chairman, CBS[30]

True Blue? In October 1994 Thomas Nicely, a math professor at Lynchburg College, Virginia, found a flaw in the way Intel's Pentium chip did division. Intel had known about the flaw since the summer but had calculated that an error would occur only once in every 9 billion calculations. The average user would have to wait 27,000 years to encounter a problem. It had decided that there was no need to alarm people.

But once Nicely posted a message on the Internet, people were alarmed. First, they were alarmed by the notion that a computer chip could actually make a math error. Then they were even more alarmed that Intel had known about the flaw but hadn't told them.

Intel argued that people were overreacting, but did offer to replace Pentium chips on a case-by-case basis.

The tide was turning in Intel's favor when in early December, just as the Christmas season was starting, IBM made the surprise announcement that the flaw in the Pentium was much more serious than Intel was letting on. According to IBM's calculations, a spreadsheet user might encounter a problem once in every 24 days – not once in every 27,000 years. To protect its customers, IBM halted shipments on its Pentium-powered machines.

Over the following weeks and months, there would be much debate over who was right – Intel or IBM.[31] In the meantime, IBM had created a critical level of uncertainty that Intel could not dispel. A week after IBM's announcement, Intel changed its position and offered a no-questions-asked return policy.

At first blush, IBM looked like the Good Samaritan, doing right by customers. But that wasn't the whole story. IBM was still pushing its 486-powered machines; Pentium-powered machines accounted for fewer than 5 percent of its sales. By contrast, most hardware makers – Acer, AST, Dell, Gateway, Packard Bell, and others – were aggressively pushing Pentium machines. It would be to IBM's benefit if nervous customers decided to play safe and continue to buy 486 machines. There was also the fact that IBM was working to develop the PowerPC chip, a rival to Intel's line. IBM wouldn't mind too much if Intel's reputation got a little tarnished.

But IBM had made its own miscalculations. Its attempt to stir up a fog backfired. Cynical observers asked whether IBM's announcement wasn't just a little self-serving. And once Intel improved its return policy, customers flocked back to the Pentium machines, leaving IBM with a large inventory of the now-less-desirable 486 machines. At this point, IBM was hardly in a position to jump on the Pentium bandwagon.

4 Is PART the Whole?

You can change the game by changing people's perceptions. This is the domain of tactics. In some sense, *everything* is a tactic. Everything you do, and everything you don't do, sends a signal. These signals shape people's perceptions of the game. And what people collectively perceive to be the game is the game. You need to take account of perceptions to really know what game you're in and to be in control of how you change it.

We've now looked at four levers of strategy: Players, Added values, Rules, and, in this chapter, Tactics. Is that it? Is PART the whole of strategy?

In principle, it is. At a fundamental level, there is only one game. Everyone interacts with everyone else, directly or indirectly, to pursue their various ends. Everything is ultimately connected to everything else. The game that includes all these interconnections might be enormous, but, in theory, that's the game. And the cast of players, added values, rules, and perceptions together would completely describe this – somewhat mythical – giant game. If it were possible to handle all the complexities of one gigantic game, then PART would be a complete set of strategic levers.

The practical reality, of course, is different. The mind divides things in order to conquer them. People draw boundaries, make compartments, add mental partitions. And they know that everyone else does, too. Everyone behaves as though there are many games, operating more or less independently.

This provides one further lever for changing games, a lever ultimately as important as each we've discussed so far: you can change the game's boundaries, alter its scope. This is the subject of our next chapter.

8 Scope

**No man is an island, entire of itself; every man is
a piece of the Continent, a part of the main.**
— John Donne, *Devotions*

No game is an island. Even so, people draw boundaries and divide the world up into many separate games. It's easy to fall into the trap of analyzing these separate games in isolation – imagining that there's no larger game. The problem is that mental boundaries aren't real boundaries – there are no real boundaries. Every game is linked to other games: a game in one place affects games elsewhere, and a game today influences games tomorrow. Even the mere anticipation of tomorrow's game influences today's.

Understanding, playing off, and changing the links between games is our fifth, and final, lever of strategy. The first step is to recognize the links between games. The links are there. Even if you don't see them, you can still trip over them, as we saw back in the Game Theory chapter when we looked at the story of Epson's entry into the laser printer market. Once you've seen the links, you can use them to your benefit. The links aren't ironclad: you can create new links between games or sever existing ones. And by doing so, you can change the scope of the game.

1 Links between Games

What types of links can exist between games? Happily, we already know the answer to this question. Here's why. We've said that there's really only one 'big' game – one game extending across space, over time, down generations. Any two games, even if conceived of as games in their own right, are really only components of the big game. By definition, this mythical big game is a game without boundaries, without a defined scope, if you like. So, PART – Players, Added values, Rules, and Tactics – describes all the elements of the mythical big game.

Since PART describes the whole, it must, in particular, describe how the pieces of the whole fit together. That is, it must describe the links between any two games, since any two games are no more than components of the big game. PART, then, is the way to classify the links between games.

Start with *Players*. Anytime there's a player in your game who's also a player in another game, the two games are potentially linked.[1] The player in common could be anyone in your Value Net – any of your customers, suppliers, competitors, or complementors. It could also be you, of course. The existence of a common player determines only the *possibility* of a link between two games. To determine whether the two games are linked – and if so, how – you have to go through the rest of PART.

Links through *Added values* can arise whenever your customers or suppliers participate in more than one market. Our discussion of complements in the Co-opetition chapter was all about this type of link. For example, Intel's and Microsoft's games are linked. For one thing, they share the same customers. But more than that, Intel raises the added value of Microsoft, and vice versa. People borrow money to buy cars, and so the games of selling cars and lending money are also linked. Ford recognized that link and decided to enter the credit game. Ford Motor Credit raises Ford's added value in the game of selling cars. The other side of this equation is when, by entering another game, you end up competing with, rather than complementing, yourself. You lower, rather than raise, your added value in the original game. This issue of cannibalization is the added-value link we'll focus on in this chapter when we present epilogues to the earlier stories of Nintendo and Softsoap.

Rules impose constraints on what players can do, and these constraints can link what would otherwise be separate games. We've already seen this effect in the Rules chapter: a most-favored-customer clause prevents a seller from treating two otherwise independent negotiations with customers as separate games. In this chapter, we'll again start in the business-to-business arena, when we look at linking games through choice of contract length. Then we'll turn to a discount-pricing rule that links games in mass consumer markets.

Finally, two games can be linked for no other reason than that someone perceives them to be linked. Thus, *Tactics*, by changing perceptions, can change the links between games. For example, issuing threats and establishing precedents are tactics that work by creating linkage across games. We'll analyze some examples of these tactics, ending with another look at our first case in the book, the game between NutraSweet and Holland Sweetener.

As this classification of links suggests, every story in this chapter could, in principle, have been told earlier in the book. But, in practice, thinking in terms of a separate *Scope* lever is very useful. It's too complicated to think of everything as one large game. Hence this chapter. The cases we've saved for here are the ones that emphasize strategies for linking games that otherwise would not naturally be linked, or for severing links between games that otherwise would naturally be. In every case, it will be taken as a given that the games have a player or players in common.

2 Links through Added Values

Newcomers to a business face many disadvantages. They lack proven products, brands, loyal customers, manufacturing experience, and relationships with suppliers. As a challenger, if you go head-to-head with an incumbent, you're likely to lose. There's little you can do that the incumbent can't do as well, if not better. In short, you have little added value.

But you don't have to go head-to-head. Instead, you may be able to play off the links between the business you're targeting and the incumbent's existing business. You do something that the incumbent can't match without hurting his existing business. You create a dilemma for the incumbent. He wants to – and *could* – come after you, but he doesn't. That's because if he did, he'd lower his added value in the game he's already playing – and that cost would be too high. So, for now at least, the incumbent leaves you alone.

Playing judo

The Japanese art of judo teaches how to use an opponent's weight against him, to turn his strength into weakness. In business, the judo strategy exploits links between games to turn the incumbent's strength into a handicap. Judo explains how Sega was able to topple the video game giant Nintendo.[2]

I don't like the idea of one company monopolizing an industry.
— Hayao Nakayama, President, Sega Enterprises[3]

Super Sonic In the Added Values chapter, we left Nintendo with a stock market value exceeding Sony's or Nissan's, and with Mario being better known than Mickey Mouse. That was in 1990. Three years later, Mario was still more popular than Mickey Mouse among US children. But now there was an even more popular icon: Sonic the Hedgehog.[4] Did Sonic just appear out of the blue?

Sonic the Hedgehog was the creation of Sega, a rival video game manufacturer. Unable to establish more than a toehold in the 8-bit game, Sega didn't give up. Instead, it developed a faster, more powerful 16-bit system. It took Nintendo two years to respond with its own 16-bit machine, but by then, with the help of Sonic, Sega was already well on its way to establishing a secure and significant market position.

Was it mere luck that gave Sega such a long, uncontested period to establish itself in the 16-bit game? Was Nintendo simply asleep at the wheel? The answer is more complicated than that, and this is where the link between games comes in. The new 16-bit game and the old 8-bit game were closely linked. When Sega launched the 16-bit game, Nintendo's 8-bit franchise was at the height of its value. That gave Sega a judo opportunity to use Nintendo's strength against itself.

Origins: Service Games The name Sega, although it sounds quintessentially Japanese, is actually short for 'Service Games.' The company was founded in Tokyo in 1951 by two ex-servicemen to import arcade games and jukeboxes to US military bases in Japan. A few years later, David Rosen, another American ex-serviceman, began importing amusement machines to Japan. The two operations merged in 1965 and became Sega Enterprises. Soon afterward, Sega launched Periscope, a 'torpedo' game, which turned out to be a big success in Japan. Sega brought Periscope to the United States, where it broke the dime price barrier to become the first 25-cent arcade game. Sega's success caught the attention of conglomerate Gulf+Western, which acquired the company in 1969.

Sega used its knowledge of arcade games to move into home video games, introducing an 8-bit system, the SG-1000, in 1983. But the SG-1000 never really took off, selling fewer than 2 million units in Japan and the United States. Sega couldn't break into Nintendo's virtuous circle.

Gulf+Western lost interest in Sega and in 1984 sold the company to a management buyout team. David Rosen, Hayao Nakayama (president of Sega in Japan), and Japanese software house CSK joined forces to buy Sega for $38 million.

The New Beginning: Genesis In October 1988 the reborn Sega introduced its 16-bit Mega Drive home video game system. Based on Sega's arcade machines, the 21,000-yen ($165) Mega Drive had many advantages over 8-bit systems, including better sound, more colors, and the capability to display multilayered images.

Sega adapted a number of its arcade games for the Mega Drive but found it hard to sign up third-party software developers to design games for the system. During its first year on the market, the Mega Drive chalked up sales of only 200,000 units.

Sega's 16-bit system came to the US market in September 1989. Renamed Genesis, the Sega system sold for $190, and games were priced between $40 and $70. Sega came up with some hit games but no home runs. One title, *Altered Beast*, based on Sega's line of arcade games, gained notoriety for its graphic violence. Pop star Michael Jackson helped Sega develop *Moonwalker*, a game in which Jackson used his dancing skills to overcome attackers. Sega also built up alternative distribution channels, selling through software retailers Electronics Boutique, Babbages, and Software ETC. Still, sales were slow.

Sega's fortunes started looking up in 1990, when Nakayama hired Tom Kalinske, formerly of Matchbox, to head up Sega's US operations. Kalinske realized: 'We have got to reduce the price of the Sega Genesis by $50. We've got to bundle in our best software – Sonic the Hedgehog – and advertise to the world that we are better than the competition.'[5] In June 1991 the Genesis, together with Sonic, went on sale for $150. Promoted with the slogan 'Genesis does what Nintendon't,' the Sega system became the 'cool' machine to have. Sales soared, and software developers rushed to turn out games for the system.

Super Nintendo Nintendo had been developing a 16-bit video game system since the late 1980s but was in no rush to bring it to market. Before Sega entered, Nintendo was content to grow the 8-bit base and thereby grow the potential customer base for its future 16-bit machines. Nintendo's thinking was that an 8-bit system today complements a 16-bit system tomorrow. By waiting, it could sell its 8-bit system today and its 16-bit system tomorrow, instead of just a 16-bit system today. As Bill White, Nintendo's US public relations director, explained: 'The Nintendo philosophy is that we haven't maxed out the 8-bit system yet.'[6]

If Nintendo jumped too quickly into the 16-bit game, there was even a risk that software houses and retailers would abandon the 8-bit market. Software houses might turn to 16-bit games and develop fewer new 8-bit titles. Retailers might cut the shelf space allocated to 8-bit games and discount the 8-bit titles they did continue to carry. The 8-bit market was Nintendo's golden goose. Why risk killing it?

After Sega introduced its 16-bit system, these arguments might have seemed moot. Yet Nintendo was still in no hurry; consumers would wait for their 16-bit systems. A year after Sega introduced the Mega Drive, Nintendo gave Japanese consumers a reason to wait a little longer. It pre-announced its 16-bit Super Famicom system in late 1989 and began shipping the machines about a year later. *Super Mario World*, a 16-bit game developed by Nintendo's ace designer, Sigeru Miyamoto, was released at the same time. Within five months, sales of the 32,000-yen ($200) Super Famicom overtook competing 16-bit systems. Nintendo President Hiroshi Yamauchi, expressing no surprise at Super Famicom's success, reiterated his view that 'the name of the game is the games.'[7]

Nintendo wasn't in a hurry to jump into the 16-bit game in the United States, either. Following the pattern established in Japan, it wasn't until September 1991, two years after Sega had introduced Genesis, that Nintendo introduced its 16-bit system to the United States. Renamed the Super Nintendo Entertainment System (Super NES), the machine was priced at $200, and games cost between $50 and $80.

The 16-Bit Battle When Nintendo finally entered the 16-bit market, the competition between Sega and Nintendo to sell 16-bit systems became intense. They competed with price cuts, threw in free software, and raced to put together the longer list of game titles. By the end of 1991 Genesis had 125 game titles, while Super NES had only 25. Nintendo claimed that close to 2 million Super NES machines had been sold. But Sega argued that 1 million, at most, was a more accurate figure and that it was Genesis sales that had truly topped 2 million. In May 1992 Nintendo reduced the price of the Super NES to match Sega's $150 price. Nintendo claimed to have a 60 percent share of the 16-bit market, while Sega boasted a 63 percent share! The same month, competition heated up even further when Nintendo and Sega each brought out stripped-down versions of their 16-bit machines for under $100.

Delaying the 16-bit battle was a very good reason for Nintendo not to rush into the 16-bit game. Once Nintendo jumped into the game, there was competition and, hence, much lower 16-bit prices. That shrank the 8-bit pie – and Nintendo's added value in the 8-bit game along with it. The price of 8-bit cartridges came down $20, and only seventy-five new 8-bit titles came onto the market in 1991 – fewer than half the number in previous years. The 8-bit pie wasn't gone, but it was half-eaten.[8]

For as long as Nintendo let Sega have the 16-bit market all to itself, 16-bit prices remained high. These high prices cushioned the effect the new technology had on the added value of the old. By staying out of Sega's way, Nintendo made a calculated trade-off: give up a piece of the 16-bit pie in order to extend the life of the 8-bit one. Nintendo's decision to hold back was reasonable, given the link between 8-bit and 16-bit games.

Still, it was a hard call. By delaying, Nintendo granted Sega a temporary monopoly of the US 16-bit market, and Sega was able to get something close to a virtuous circle rolling. It took until September 1994 – three years of playing catch-up – before Nintendo managed to overtake Sega in the 16-bit game.

Professor Dorothy Leonard-Barton of Harvard Business School explains that an organization's core competencies in one generation of technology can turn into 'core rigidities' as far as the next is concerned.[9] That's one reason that established players often find it hard to make the transition to the next-generation technology, and that's one reason why technological change often gives challengers opportunities to overturn incumbents. The story of how Sega got a window of opportunity is different, though. Nintendo had the 16-bit technology but deliberately chose to delay entering the 16-bit game.

Many people have criticized Nintendo for its delay, saying that it would have been understandable if no one else had yet introduced a 16-bit system. But once Sega brought out the Genesis, Nintendo should have been hot on its heels with its own 16-bit entry. In delaying, Nintendo gave away the farm – to a hedgehog. It apparently forgot the adage that it's better to eat your own lunch than to have someone else eat it for you. Cannibalize yourself rather than let someone else eat you alive.

We think the decision to delay wasn't obviously an error. Nintendo faced the classic dilemma that most successful companies eventually face. You've come up with a great product and you dominate a market, but then a challenger comes along with a new and superior technology. As long as the challenger has a monopoly on the new technology, it has an incentive to charge a high price. That limits the pace of adoption – which is good news for you, since that extends the life of your product. Once you jump into the new technology, you force the challenger to compete head-to-head with you. The price of the new technology will fall and, along with it, the added value of your old product.[10] Epson learned this lesson the hard way. Nintendo was more careful. While you can't wait forever to make the transition, that doesn't mean you should jump right in.

As for Sega, it parlayed Nintendo's 8-bit strength into a 16-bit weakness – but only because it didn't decimate the 8-bit market. Had Sega priced its 16-bit system to compete with 8-bit systems, Nintendo wouldn't have faced a dilemma. Nintendo would have had nothing to lose by jumping quickly into the 16-bit game, and that would have made life much harder for Sega. The judo strategy is based on the idea that a challenger has nothing to gain if the incumbent has nothing to lose.

The story of Sega and Nintendo in 16-bit video games shows how a challenger can create a window of opportunity for itself by turning the incumbent's strength into weakness. Even if that's not possible, a challenger may still be able to use the judo approach to at least neutralize the incumbent's advantage.

Fear of Failure In the Added Values chapter, we told the story of entrepreneur Robert Taylor and his innovative Softsoap liquid-soap product. Taylor's problem was how to prevent the likes of Procter & Gamble and Lever Brothers from copying his idea. He got a window of opportunity when the majors adopted a wait-and-see stance.[11]

Why did the majors play wait-and-see? Early on, the success of liquid soap was far from assured. Although liquid soap was convenient and eliminated the puddle of soap ooze left behind by bar soaps, it wasn't a technical breakthrough. There was no compelling reason for people to switch from bar soap.

Given the uncertainty over the liquid-soap concept, it made sense for the majors to sit back, save their money, and hope that Taylor failed. For a major, jumping into liquid soap would only help validate a category that offered it very little upside. Liquid soap was unlikely to expand the overall soap market. If successful, sales of liquid soap would surely come almost entirely at the expense of bar soap.[12]

The majors probably assumed that if Taylor did start taking share away from bar soap, they could deal with the problem when it arose. At that point, they would do better to cannibalize their barsoap sales themselves rather than let Taylor do it. That's the idea, again, of eating your own lunch rather than letting someone else eat it. But the majors figured that they could wait until that point to bring out competing liquid-soap products and still be able to take back most of their lost share. After all, they had the distribution channels and the brand names. So they waited and watched.

After Taylor had repeated his success in test markets with a strong national rollout of Softsoap, the majors felt it was time to do their own tests. That was when they discovered the shortage of pumps. Remember: Taylor had bought 100 million pumps, locking up a year's supply.

Once the majors were over this hurdle, they found themselves facing a hard decision: should they use their bar-soap brands on the new liquid-soap products? For example, should P&G launch a liquid-soap product under its flagship Ivory brand or under some new name? The Ivory name would significantly increase the chances of success, but it would also create a closer link between the liquid-soap and bar-soap games. If P&G's liquid-soap product failed, its immensely profitable bar-soap business would be damaged.

Procter & Gamble had several good reasons to be nervous about creating a closer link between the two games. First, liquid soap wasn't really a 'soap' at all. It was a detergent-based product – an entirely different chemical formulation. Consumers had a very conservative attitude toward soap and, historically, had resisted detergent-based cleansing bars.[13] Second, liquid soap had traditionally been used only in institutional settings, where it was essential for reasons of hygiene. Could liquid soap develop a different, more upscale image for the home? If consumers disliked Ivory's liquid-soap formulation, or if they associated it with grimy institutional settings, P&G's valuable Ivory brand name would be sullied.

The uncertainties surrounding liquid soap temporarily neutralized the major players' advantages. Because they didn't want to jeopardize the large added values of their brand names, the majors decided to play the liquid-soap game independently of the barsoap game. They entered the liquid-soap category without using their bar-soap brand names. Procter & Gamble test-marketed a liquid soap named Rejoice. The test was a failure, and Procter & Gamble again delayed. Armour-Dial, too, decided not to risk its Dial brand name on its liquid-soap product. Its entry, Liqua 4, which sounded more like a drain opener than a soap, also failed in the marketplace. No doubt the Dial brand name would have worked a lot better. After three years, Procter & Gamble finally rolled out Ivory brand liquid soap, with extremely aggressive pricing, trade promotion, couponing, and advertising, in addition to the trusted brand name. Liquid Ivory was a success, capturing 30 percent of the market.

With all the delays by the majors, Taylor had a sufficient window of opportunity to establish some permanent brand loyalty for Softsoap. Even after the arrival of Liquid Ivory, Softsoap maintained its lead position.

The Softsoap story shows that, perhaps paradoxically, you can do better if your product has some chance of failing. With a lot to lose from jumping full force into a new product that could fail, established players are likely to hold back. Failure damages the added value of their other products, especially if the new product is clearly identified in any way with their existing ones. For that reason, they may avoid using existing brand names. They're not going to want to link the proven and unproven games too closely, so until the new product is proven successful, they'll remain cautious. All of this uncertainty neutralizes what would otherwise be a major advantage to the established players.

Uncertainty can be the challenger's friend. Not too much uncertainty, of course, but not too little, either. The trick is to have a moderate level of uncertainty, enough to induce incumbents to hold back, play cautiously, and keep the games apart.

Summing up our two stories in this section – Sega and Softsoap – both challengers used judo strategies to create a window of opportunity. Sega benefited from Nintendo's reluctance to kill the 8-bit video game market, while Softsoap benefited from Procter & Gamble's reluctance to risk its Ivory brand.

PLAYING JUDO

How can a challenger counter an incumbent's strength?

THE STRATEGIES

1. A challenger prices a superior product sufficiently high to avoid eating into sale of the incumbent's existing product.
2. A challenger bets on an unproven product – one with some chance of failing.

WHY THEY WORK

1. The incumbent holds back from copying the challenger – copying would trigger price competition and accelerate cannibalization of its existing product.
2. The incumbent may copy the challenger – but doesn't apply its existing brand names for fear they'll be damaged if the product fails.

In both cases, the incumbent faces a dilemma.

The old drives out the new?

The story of 16-bit video games is a classic example of how established players delay bringing out the next generation of products in order to limit competition with the current-generation product. Even after they decide to bring out the next generation, they aren't off the hook. Just as the new competes with the old, the old competes with the new. The old product doesn't just disappear. So when you do bring out a new product, the challenge is to prevent the loss of added value that results from competing with your old product. To protect your added value, do what you can to sever the links between yesterday's and today's games.

Revise or Perish It may be hard to believe, but some college textbooks are updated every year. Of course, textbooks get outdated and need to be revised but probably not that often. Why do publishers produce such frequent revisions of textbooks?

What's really going on is that a used textbook is an extremely good substitute for a new one, especially when new textbooks cost $50 or more. Former students would be more than happy to recoup some of their $50 investments by selling their used books to the next cohort of students, who'd be happy to save the money. Students win and publishers lose.

This year's game of selling textbooks is linked to last year's game. Last year's new textbooks become this year's used ones and undermine the added value of this year's new texts. To maintain the demand for new textbooks, publishers have to find a way to get used books off the market. They want to sever the link between last year's game and this year's game.

The publishers' strategy is to revise, and revise again. Instructors prefer to teach their courses from the latest version of the textbook, making all the previous versions obsolete. The new students now have a big incentive to buy the new book, so the large stock of used books now has less effect on the added value of the new ones.

Frequent revisions work, but they're costly. There are some other strategies publishers might want to try. They could rent textbooks along with selling them: say, $60 to buy and $30 to rent. This way, students wouldn't buy a textbook unless they planned to keep it, and the resale problem would be solved. In the future, electronic publishing will give publishers even more control over their material. They'll be able to 'rent' books by providing network access to the material over the length of the course. That won't mean that frequent revisions will become a thing of the past. On the contrary, publishers will be able to revise electronic books on an almost continuous basis, but for content reasons alone.[14]

Publishers aren't the only ones who play the revision game. Software makers are masters at it. Clothing designers, record companies, fragrance makers, food magazines, and automakers play the revision game, too. This year's new look, sound, smell, taste, and feel make last year's obsolete. People have to keep buying in order to keep up.

But you can get carried away with the obsolescence strategy. To see how you can go wrong, let's return to the story of 16-bit video games.

Nintendo's engineers had to decide whether to make their new 16-bit system capable of playing Nintendo's existing 8-bit games. They decided against backward compatibility. One reason was to keep down the cost of the new hardware.[15] But there was another reason for backward incompatibility. If 8-bit games had worked on 16-bit machines, that would have lowered the added value of all the new 16-bit games. Backward incompatibility made obsolete all those 8-bit games that kids might already have had or that their friends might have lent them. Just as publishers force students to buy the latest version of a textbook, Nintendo forced its 16-bit customers to stock up on new software.

Nintendo may have been a bit too clever. It didn't have the 16-bit game to itself. Nintendo found itself behind Sega in the race to build a 16-bit customer base. But it did have a potential advantage. Since Nintendo owned the 8-bit market, there was a massive stock of Nintendo 8-bit software out there. Backward compatibility would have been a strong selling point for Nintendo. One industry analyst said that 'Nintendo will be going back to square one when it introduces its 16-bit system . . . and it will be competing evenly with Sega. . . .'[16] Not so had Nintendo made its 16-bit systems backward-compatible. By making its 16-bit system incompatible with its 8-bit system, Nintendo needlessly leveled the 16-bit playing field.

3 Links through Rules

Rules are a direct lever for changing the scope of the game. We'll look at two examples in this section – one applicable in a business-to-business setting and one relevant to mass consumer markets.

In business-to-business dealings, you can control the scope of the game by the length of the contracts you write with your customers and suppliers. One-year contracts create a series of one-year games, whereas a five-year contract turns five one-year games into one five-year game. If you have the power, you can choose the length of the game you – and others – play.

The Long and the Short of It Two suppliers are competing for your business – your incumbent supplier and a challenger. The two suppliers are pretty much alike, and each can meet all of your needs. Thus neither one has much, if any, added value; you have it all. What should you do with this power?

Start by making some rules of the game. Do you want the game to include annual competitions for one-year contracts, or would you do better to have infrequent competitions for long-term contracts?

If you award short-term contracts, you might be disappointed by the results. When suppliers play lots of little games, they may not compete that aggressively with each other. After all, the prize is only a one-year contract – not big enough to tempt the challenger into making an aggressive bid that could have repercussions elsewhere. Remember the Eight Hidden Costs of Bidding: it's quite likely that the roles of incumbent and challenger will be reversed elsewhere, and the challenger won't want to provoke the incumbent for a small prize that he doesn't stand that big a chance of winning, anyway. For the incumbent, losing the competition isn't a disaster. There's always another opportunity next year, or other opportunities this year. Whichever supplier wins your contract, the other can expect to win a comparable contract elsewhere. Your game isn't so big that it's make-or-break. In this environment, suppliers can settle into a live-and-let-live arrangement.

These implicit links between games don't help you. You want your suppliers to treat the current contract negotiation as if it's the only game in town. You want them to play your game all out, regardless of the fallout in the other games they're playing.

You can achieve this result by lengthening the scope of the game. The longer the contract you offer, the more the suppliers will treat the game as a once-and-for-all competition. The contract is now a big enough prize to induce the challenger to bid aggressively, regardless of the fallout that creates elsewhere. The incumbent will view losing a multiyear contract as just short of the end of the world. Next time is a very long time away.

There's another reason to proffer long-term contracts if you're in a strong position. Next year, your suppliers may have more added value than they do now. Thus, you should seize the opportunity to lock in your advantage while you have it. With a long-term contract, it's as if next year's negotiations have been concluded today, so you'll be better protected against a future shift in power.

We know just how effective it can be to shift to long-term contracts. We've sat at the other side of the table and seen this strategy used against our clients. They felt they had to be a lot more aggressive on price. They didn't like the rules, but they didn't have the power to change them.

If you're on the short side of the market, ask your customers and suppliers to compete for long-term contracts. Conversely, if you're on the long side of the market, then you want to go for short-term contracts. In sum, if you're short, go long, and if you're long, go short. Of course, you're more likely to get your way when you're on the short side – that's when you have more added value.

Finally, a word of caution. Long-term contracts are hard to write, since there are more contingencies to consider. For this reason, long-term contracts are, by necessity, incomplete, which makes it more likely that someone will try to renegotiate with you later on.

GOING LONG

If you have the power, use it to require your suppliers (or customers) to compete for long-term contracts with you.

PROS

1. Since they only have one chance, suppliers (or customers) will compete more aggressively.
2. You have the power – this is the time to use it and lock it in.

CON

1. Long-term contracts can be hard to write and hard to enforce.

After our business-to-business example of how rules link games, we now turn to a mass-consumer-market example. We'll look at a remarkably effective pricing rule: 'package discounts.' A package discount links the game of selling one product with the game of selling another. We'll examine a real case, but with some simplified numbers, to explain the rather subtle theory behind package discounts.

Discounted Value Warner Bros. owns The Fugitive and Free Willy. You might say that both movies are about escapes; other than that, though, the two movies are quite unrelated. Seeing one movie doesn't necessarily make you any more, or any less, likely to want to see the other. The two movies neither compete with nor complement each other. All they share is the Warner Bros. name. And yet there's a reason to link together the way the two movies are sold.

After the theatrical release and the premium-rental cycle, Warner Bros. was ready to sell the two videos to the mass market. What price should it charge? Let's imagine that it conducted a survey of four hundred regular video buyers, and the results revealed four equal-sized market segments:

- A hundred people would pay $20 for *The Fugitive* but had no interest in *Free Willy*.
- A hundred people had just the reverse preferences: they'd pay $20 for *Free Willy* but weren't interested in *The Fugitive*.
- A hundred people said that they'd buy both movies at $20 each.
- Finally, a hundred people said they liked both movies but weren't quite as enthusiastic; they'd pay somewhere between $15 and $20 for each of the movies, say $17.50 for argument's sake.

Warner Bros.' unit costs for videos are about $5, equally split between manufacturing the shell, cartridge, and packaging; and advertising and shipping expenses.

Based on all this information, Warner Bros. decided on a suggested retail price of $19.95 for each video, or a net price of $10.95 after taking into account the video store markup. At $19.95 each, Warner Bros. would sell a total of four hundred videos – two hundred copies each of *The Fugitive* and *Free Willy*. To reach that fourth group of customers and sell an extra two hundred videos, Warner Bros. would have to lower price to $17.50. Would that be worth it? The cost of the price cut would be $2.50 – more precisely, $2.45 – on the four hundred sales above, or $1,000. The benefit would be only $700: the profit margin of $(17.50 - 9 - 5) = $3.50 times the two hundred extra sales. So Warner Bros. chose to price both movies at $19.95, meaning that the fourth group of customers doesn't buy either one.

That's a loss – both for those people and for Warner Bros. But there is a way to sell to the fourth group and not lose money.

What the studio did was to link the two games with a pricing rule: buy *The Fugitive* and get a coupon for $5 off *Free Willy*. It could just as well have said: buy *Free Willy* and get $5 off *The Fugitive*. In fact, the deal really is: buy *The Fugitive* and *Free Willy* and get $5 off the combined price.[17]

This discount has no effect on the first and second groups of customers. They'll still buy one or the other movie at $19.95, but they won't buy both. The third group of customers gets a price break: they were already buying both movies, and now they'll save $5. That's a hit to the studio's bottom line of $5 on a hundred sales, or $500. The benefit to the studio is that it induces the fourth market segment to buy the package deal. On the retail price of $35, the studio gets $17 after the store's cut on the two cassettes. The studio's costs are $10, so it makes a margin of $7 on a hundred sales for a total of $700. Actually, though this is exactly the same benefit as before, the numbers add up this time. Making $700 while giving up $500 is the right decision.

If the studio cuts the price of each video separately by $2.50, it comes out behind; but if it cuts the price of the two videos *together* by $5, it comes out ahead. This sounds like magic. The explanation is that when it discounts the package rather than the individual items, the studio manages to get the same stimulation in demand while giving fewer people the discount. The benefit of the price break is the same, but the cost is halved.

In this example, both movies in the package were owned by one studio. But the pricing rule would work just as well if the two movies were owned by different studios. Since the combined profits on the two videos would be higher, there would always be a way to allocate the cost of the discount so that both studios came out ahead.

This principle is a very general one. There needn't be any connection between the two products or the two companies selling them. Here are some examples of cross-company couponing practices we've seen:

- Buy a Rubbermaid Servin' Saver and get 20 cents off Vlasic pickles.
- Open an account with Fleet Bank and get $100 off a Delta Airlines ticket.
- Shop at Stop & Shop and save between $50 and $100 off a Northwest Airlines ticket.
- Sign up with SNET Cellular and get AAA roadside assistance for free.
- Buy an AST Bravo notebook and get a ski package at Vail.
- Join National Car Rental's Emerald Club and get $75 off the membership price of a Diners Club Card.

Package discounts are much more widely applicable than you might ever imagine. Pick any two products at random, and it's a good bet that there's more money to be made by offering consumers an option to buy the package at a discount. Package discounts really are magic. Although they aren't rare, they aren't that common, either. This strategy has a huge untapped potential.

DISCOUNTED VALUE

Increase sales without giving up as much margin by offering package discounts.

There are just two caveats. Package discounts are least effective when the people who like one of the products also tend to be the ones who like the other. Think of *Star Wars* and *The Empire Strikes Back* rather than *Free Willy* and *The Fugitive*. In such cases, most customers either buy both products or neither. A package discount is then more like a plain old price cut and doesn't have any special effect.

The second pitfall to watch out for is resale. If a market for the coupons springs up, then the seller loses the ability to price the package. In effect, each item becomes priced individually once again. If the coupon value is small, a resale market is unlikely to arise, but this becomes a real concern if the discount is large. There are some clever ways to get around the problem. Season tickets for college football are usually sold at a discount. However, some schools give you just one ticket for the season – as opposed to a book of tickets. If you try to sell your ticket to a particular game, you may not get it back to use for the rest of the season.

4 Links through Tactics

Two games are linked anytime someone *perceives* them to be linked. The two games become one larger game when someone believes that what will happen in one of them is contingent on what happens in the other. By creating or destroying these perceptions of linkage, tactics change the boundaries of the game.

Threats and promises are the classic examples of creating a perceptual linkage. They are designed to persuade other people to do something – or not do something – based on how you say you'll respond. They do something in one game because of what they perceive you'll do somewhere else. Setting a precedent is another tactic designed to link games. You take an action in a game today to convince people of what you'll do next time you're in a similar game.

You can institute a link unilaterally. All you have to do is convince other players that you perceive two games as linked and that you're going to treat them as linked. To deal with the consequences, other players will have to treat the games as linked, too.

Creating linkage is a familiar idea: one person tries to bring in another issue, and the other person resists. The concept of linkage is often explicit in trade negotiations. What issues can be discussed together? Rice and beef imports, most-favored-nation status, human rights, military aid, copyright protection, fishing rights. The links can be very tight or very loose: an explicit threat or promise to act in a certain way, or a vague hint of a general policy.

The key ingredient for linkage is contingency: another player must believe that what you will do in one game depends on what happens in another. Or you believe that he believes this. Or he believes you believe that he believes this. The game is all about perceptions.

Cable Retransmission Dissent In the Tactics chapter, we took a first look at the negotiations between the broadcasters and the cable TV companies over cable retransmission consent. Most broadcasters — CBS, in particular — lost out. Here, we'll take a look at the game in southern Texas, where things turned out a little different.

Unlike broadcasters elsewhere, the Corpus Christi stations didn't allow themselves to be painted as the bad guys. They took the lead in shaping public opinion. Mike McKinnon, owner and founder of KIII, the local ABC affiliate in Corpus Christi, used his station to broadcast the following challenge: 'If cable will reduce your basic rate by 60 cents per subscriber, we'll give them permission to carry our total programming lineup at no charge from us. Free means free.'[18] TCI, the local cable operator, was charging subscribers $10.23 a month for basic cable — 60 cents a channel. Yet it wasn't willing to pay the broadcasters anything for their signals. If KIII wasn't going to get paid, neither should TCI.

Many things are done a little differently down in Texas. In most of the country, television stations are regularly bought and sold, but all three network affiliates in Corpus Christi were owned by their original founders. The CBS affiliate KZTV had been founded in 1956 by the now eighty-eight-year-old Vann Kennedy. The NBC affiliate KRIS had been founded by one T. Frank Smith, Jr., in the same year. The three owners – McKinnon, Kennedy, and Smith – went way back. TCI could try to divide and conquer, but the three owners trusted each other to hold the line. They knew full well that if one of them struck a deal with TCI, the other two might have to cave in without getting paid. If all three of them held out, TCI would ultimately pay for their programming.

When the negotiation deadline passed without an agreement, all three stations went 'off the cable.' Viewers were furious that their cable company wasn't carrying any of the three networks. TCI tried to appease them by handing out forty thousand A/B switches that allowed their customers to switch back and forth between cable and over-the-air reception. But this Band-Aid wasn't a real solution.

TCI's customers were unhappy, and TCI wasn't the only game in town. Omnivision, a local 'wireless cable' operator, had paid for retransmission consent from all three affiliates. The stations were off TCI, but they were still on Omnivision. As a result, customers started flocking to Omnivision – over two thousand in the first two weeks alone.

TCI decided to play hardball. It found an opportunity in nearby Beaumont, Texas, where McKinnon, owner of KIII in Corpus Christi, was in a more vulnerable position. In Beaumont, McKinnon owned the local ABC affiliate, KBMT. The other two Beaumont affiliates had given retransmission consent for free, and since McKinnon couldn't afford to be the only one off-cable, he, too, offered TCI his Beaumont signal for free. He was in for a surprise. TCI turned him down.

TCI linked the game in Corpus Christi to the one in Beaumont: if McKinnon wanted to get KBMT back on-cable in Beaumont, then his KIII in Corpus Christi would have to stop holding out. As McKinnon explained: 'We were ready to give them consent in Beaumont, but [TCI Senior Vice-President Robert] Thompson said in a press conference that they were holding KBMT hostage for Corpus Christi.'[19] A TCI spokesman confirmed that this was indeed their game plan: 'If [McKinnon] gets his way in Beaumont, he'll get it elsewhere as well.'[20]

McKinnon was in an impossible position. Being off-cable in

Corpus Christi was a problem, but the other two network affiliates were off, too, so at least he wasn't losing viewers to them. In Beaumont, though, he was the only one off-cable, and his viewers were switching to the rival networks.

Kennedy and Smith, McKinnon's fellow broadcasters in Corpus Christi, understood TCI's tactic and recognized that McKinnon's problem was their problem, too. It wasn't in their interest to have McKinnon over a barrel in Beaumont. If TCI could get him to give away consent in Corpus Christi, then their bargaining position would be seriously compromised. So Kennedy and Smith came up with their own linkage. They refused to discuss any TCI proposals in Corpus Christi until McKinnon was back on-cable in Beaumont. 'They are holding [KBMT] hostage. We can't do much until they put them back on the air,' commented Smith.[21] The tactic worked: McKinnon was put back on-cable in Beaumont, and the link between the Beaumont and Corpus Christi games was severed.

Although TCI accepted that it would have to pay something to the broadcasters in Corpus Christi, it still worried about setting a precedent. TCI didn't want to create the perception that it was willing to pay for retransmission consent elsewhere. It didn't want any future negotiations to be linked to what happened in Corpus Christi.

Smith suggested there might be room for TCI to be creative. As he put it: 'I don't give a damn who gets the money, but if they have my [KRIS] signal on cable, somebody is going to get paid for it.'[22] That somebody turned out to be the Corpus Christi campus of Texas A&M University, to which TCI donated an undisclosed amount of scholarship money. In return, the three broadcasters gave TCI retransmission consent.[23] TCI's oblique method of paying in Corpus Christi managed to preserve some uncertainty over whether it would really pay for retransmission consent elsewhere.

Some games could be quite naturally linked but are better left unlinked. The following disguised case shows that sometimes it's better not to expand the scope of the game.

Don't Mention It Melanie, the CEO of a large textile maker, was pleasantly surprised when one of her larger customers called and asked to increase the size of the current year's order. Back in January,

Melanie had contracted to supply the customer over the course of the year at a fixed price. In return for the price guarantee, the customer had agreed to buy exclusively from Melanie.

The good news was that the customer's business was doing so well that he could see he was going to need more product before year-end. The bad news was that the customer wanted a discount for the extra volume — a 10 percent discount. Melanie said that she'd get back to him.

Melanie's CFO was dead set against giving the discount. The customer had agreed to a price and was now trying to wriggle out of it. Giving the discount would not only be throwing away money today, it would set a bad precedent. The customer would get the idea that future contracts could be renegotiated. Even worse, the discounted price might become a new baseline for future contract negotiations.

Rather than just give away money, the CFO wanted to link the discount to a contract extension. The customer might tell Melanie that the discount would make him favorably disposed come contract-renewal time, but why leave things to chance? The CFO suggested to Melanie that there be a quid pro quo: the customer gets the discount and, in return, agrees to extend the current contract for another year.

Melanie saw this as a very dangerous tactic. It was only July, and the current contract still had five months to go. In the normal course of events, she wouldn't even start to discuss next year's contract until November. Asking for the quid pro quo would open the subject today, and that was the last thing she wanted to do.

Melanie knew that November was the right time to negotiate a new contract. At that late date, the customer would be hardpressed to find a replacement supplier who could deliver in January. However, if Melanie started negotiating next year's contract now, the customer would have several months in which to qualify a replacement. That's why she thought it best to keep discussions about the discount and the next contract as far apart as possible.

Melanie had another reason to keep the two games separate. Quite apart from whether it was wise to put next year's contract on the table now, she was concerned that the customer might resent the linkage.

The CFO accepted Melanie's arguments but countered that giving in to the customer would make them look like pussycats. But Melanie looked at it differently: she didn't mind if the customer perceived her to be a pushover. Come November, the customer would likely be a little more confident than usual going into the annual contract negotiation and therefore a little less prepared.

Melanie called the customer back and reiterated that the current contract required him to buy all his product from her at the current price. That said, she was happy that his business was doing so well and wanted to help out. What discount did he really need? The customer said 7 percent, and it was a done deal.

As it turned out, the year-end contract negotiations were quite tough and dragged on well into December. Finally, the customer threatened to drop Melanie if she didn't agree to his terms. Melanie told the customer that he might want to consider how he would get supplied next month. The customer realized that at this late juncture he didn't really have any alternatives to Melanie. His bluff had been called. Melanie signed the customer to a new, multiyear contract on terms that both of them could live with. The base price would remain the same as last year's, and incremental volume over last year's total would be rewarded with discounts.

Our last case study in the book brings us full circle. We began our exploration of PARTS with the story of Holland Sweetener's entry into the US aspartame market once NutraSweet's patent expired. We saw how Holland helped save Coke and Pepsi hundreds of millions of dollars without getting much in return. That was the endgame. Here, we return to take a look at the opening game as it played out in Europe.[24]

Sweet Temptations NutraSweet's European patent expired in 1987, five years before the US patent expired. Although the European market for aspartame was small, and not especially valuable in its own right, the game in Europe was an important harbinger of things to come in the United States.

Holland Sweetener entered the European market with a small, 500-tonne plant. There was an element of judo in Holland's strategy. NutraSweet, the established player in Europe, owned the market. Cutting price to kill off the challenger would hurt NutraSweet much more than ceding some share. And yet, shortly after Holland's entry, NutraSweet cut price aggressively, triggering a price war. By early 1990, aspartame prices in Europe had fallen from $70 to $22–$30 per pound, and Holland was losing money.

What was NutraSweet up to? At face value, the price cuts didn't make economic sense. In slashing price, NutraSweet gave up 80 percent of its profit margins. It would have done better to live and let live and give up some share.

But looking only at the European market is too narrow a scope. NutraSweet was looking ahead to the time when the US market would open up. Holland's 500 tonnes would serve 5 percent of the world market. That wasn't the problem. The problem was that if Holland made money, it would have a natural temptation to expand. NutraSweet hoped to nip the 'tulip' in the bud.

By fighting in Europe, NutraSweet denied Holland access to the learning curve and starved it of profits. Even more important, Nutra-Sweet's aggressive response set a precedent. Its tactic was designed to create the perception that the same hostile reaction would await any other entrants into the market. A price war in Europe served as a warning to anyone considering entry into the US market after 1992. NutraSweet was hoping, no doubt, that Holland would be among those who got the message.

A message was sent, but how should it have been received? Perhaps NutraSweet was simply bluffing. Just because it had fought a price war in Europe didn't mean it was committed to fighting one in the United States. The rationale for NutraSweet's fighting in Europe was to deter entry into the US market. But if deterrence failed, and someone did enter the US market, the rationale for launching a second price war didn't exist. At that point, NutraSweet wouldn't have any other markets left to protect.

Putting itself in NutraSweet's shoes, Holland could figure out that NutraSweet had little to gain by starting a price war in the US market. In fact, it had plenty to lose. The US market, at ten times the size of the European market, seemed far too profitable to sacrifice in a price war.

By this logic, Holland could safely ignore the price war in Europe, expand capacity, and enter the US market. But taking this logic one step further, why, then, had NutraSweet bothered to fight in Europe?

Perhaps NutraSweet wasn't so cool and calculating, after all. Maybe it would launch a price war in the US, anyway. Or perhaps Nutra-Sweet *was* cool and calculating but thought it could convince Holland that it wasn't.

Bluff or not, NutraSweet's tactic worked. The price war in Europe made it harder for Holland to justify expanding capacity. It was losing money in Europe, and it now perceived prospects in the US as less rosy than it had initially thought. As a result, Holland delayed its expansion plans, and when the US market opened up to competition in late 1992, Holland could stage only a limited presence there.[25] True, Coke and Pepsi were able to use even Holland's limited presence to negotiate lower prices from NutraSweet, but those new prices would have been even lower had Holland been a more credible player. With its small capacity, Holland was unable to supply all of Coke's or Pepsi's needs. That limited the extent to which Coke and Pepsi could use Holland as a bargaining chip against NutraSweet.

Holland's small plant didn't work as a judo strategy. On the contrary, by starting small, Holland tempted NutraSweet into launching a price war in Europe. Perhaps the war *was* only a bluff, but even a small chance of the bluff working was incentive enough for NutraSweet. The cost to NutraSweet of a price war in the small European market was insignificant next to the benefits of deterring, or even just delaying, competition in the US market. NutraSweet had every incentive to create some fear and uncertainty, with the hope of dissuading Holland from expanding.

When you enter a market with only a small capacity, you run the risk that the incumbent will try to bump you out of the game. Instead, consider what might be called a 'sumo' strategy. If you plan to get big, start out big. Build a large plant at the outset. That way, the incumbent isn't tempted to respond aggressively in an effort to dissuade you from expanding later on.

5 The Larger Game

The most important lesson of this chapter is that every game takes place in a larger context. This is what allows a game's boundaries to be expanded or simply moved. Even when a player seems to be narrowing the scope of a game, it's the player's power in some bigger game that makes this maneuver possible. You may think you know what game you're playing, but that game is invariably part of a larger one. That's a good message to end with.

There's always a LARGER game.

9 Being Ready for Change

You now have all the tools you need to apply game theory to business. But, as you've probably realized, this is just the beginning. What's next?

Game theory is a tool to be incorporated into your way of thinking. Plato said that the unexamined life is not worth living. In business, you might say that the unexamined game is not worth playing.

Once you start considering what you're currently doing from the perspective of game theory, you'll stop taking many features of your business for granted. You'll realize you don't have to accept the game you find yourself in. This realization is itself extremely liberating. It allows you to see beyond the constraints of your immediate situation and frees you to seek the greater rewards that can come from changing the game.

You'll probably spot some ways to change the game to your benefit almost immediately. Make these changes and you'll find your efforts to apply game theory amply rewarded. But you're not done yet.

Changing the game is not something you'll want to do once and then forget about. It's best viewed as an ongoing process. No matter how successfully you've seized your current opportunities, new ones will appear that can be best utilized by changing the game again. No matter how secure your current position, challenges will arise that can be best met by changing the game further.

After all, other players will be trying to change the game, too. Sometimes their changes will work to your benefit and sometimes not. You may need to respond to these changes by changing the game again. There is no end to the game of changing the game.

A checklist for change

To help you become more effective at changing the game, we've made up a list of 'self-diagnostic' questions. These questions, organized according to the PARTS model, recap much of the material you've read in this book.

Players Questions

- Have you written out the Value Net for your organization, taking care to make the list of players as complete as possible?
- What are the opportunities for cooperation and competition in your relationships with your customers and suppliers, competitors and complementors?
- Would you like to change the cast of players? In particular, what new players would you like to bring into the game?
- Who stands to gain if you become a player in a game? Who stands to lose?

Added Values Questions

- What is your added value?
- How can you increase your added value? In particular, can you create loyal customers and suppliers?
- What are the added values of the other players in the game? Is it in your interest to limit their added values?

Rules Questions

- Which rules are helping you? Which are hurting you?
- What new rules would you like to have? In particular, what contracts do you want to write with your customers and suppliers?
- Do you have the power to make these rules? Does someone else have the power to overturn them?

Tactics Questions

- How do other players perceive the game? How do these perceptions affect the play of the game?
- Which perceptions would you like to preserve? Which perceptions would you like to change?
- Do you want the game to be transparent or opaque?

Scope Questions

- What is the current scope of the game? Do you want to change it?
- Do you want to link the current game to other games?
- Do you want to delink the current game from other games?

The more you ask yourself these questions, not just casually, but in an orderly, disciplined way, the more opportunities you'll find for improving your game. It's important to think methodically about changing the game. That's the great strength of the game theory approach: it helps you to see the whole game.

What you don't see, you can't change. By identifying all the players and all the interdependencies, game theory expands your repertoire of strategies for changing the game. At the same time, it helps you to evaluate each proposed change more exhaustively and, hence, more reliably. It encourages you to try out the perspectives of other players in order to understand how they'll respond to your new strategies. Out of this more comprehensive vision comes a set of strategies that is richer and more dependable.

The bigger 'bigger picture'

Our aim in writing this book was to paint a more complete picture of business relationships. Co-opetition is in the air, and we want to encourage this shift away from the focus on competition that has dominated much of business strategy. In particular, we want to counter the embattled mind-set that can cause players to miss opportunities to expand the pie.

Finding a better game to play doesn't have to come at the expense of others. This perspective makes it easier to find the best strategies, whether cooperative or competitive. In some of the cases we have discussed in this book, defeating others was the best strategy and the result was win-lose. Sometimes, that is the answer. But we don't want to presume it's the only answer before we start. Often, the best strategy has mulitple winners. In this book, we saw many instances of companies finding ways to capture a bigger slice of pie by growing the entire pie. Looking for ways to expand the pie, while keeping an eye on capturing the pie, helps promote a benevolent attitude toward other players, while at the same time keeping you tough-minded and protective of your own interests.

Business is cooperation when it comes to creating a pie and competition when it comes to dividing it up. This duality can easily make business relationships feel paradoxical. But learning to be comfortable with this duality is the key to success.

We want this book to change the game of business. By suggesting ways to make the pie bigger, we hope to make business both more profitable and more satisfying. By suggesting ways to change the game, we hope to keep business dynamic and forward-looking. By challenging the status quo, we say things can be done differently — and better. And that is our challenge to you.

Notes

1. War and Peace

1. Drawing on the work of Anne Hollander, Eric Nash points out that the clothing people wear to work also comes from war: the tie was long called a cravat, after the seventeenth-century Croatian mercenaries who wore them on French battlefields; the vestigial brass loops on trench coats are actually grenade hooks; the tailored suit can be traced back to the linen padding worn under a suit of armor; men's coats unbutton from the left so that a right-handed man might draw a sword or gun quickly. See Style column, *New York Times Magazine*, July 30, 1995, p. 39. Also see Anne Hollander, *Sex and Suits* (New York: Knopf, 1994).

2. This is a commonly quoted statistic. See the *Seattle Times*, April 24, 1994, p. A1, and *Inc.*, April 1994, p. 52. It was (successfully) used in testimony to the US Congress subcommittee on aviation to help the airlines get a two-year exemption from the fuel tax (see FDCH Congressional Hearings Summaries, March 22, 1995). The net income for the US airline industry in those years (*US News & World Report*, September 25, 1995, with data coming from Air Transport Association and PaineWebber) was:

1990	$3.9-billion loss
1991	$1.9-billion loss
1992	$4.8-billion loss
1993	$2.1-billion loss

3. *Electronic Business Buyer*, December 1993. The portmanteau 'co-opetition' was coined by Ray Noorda.

4. This is the title of an actual book: Wess Roberts, *Leadership Secrets of Attila the Hun* (New York: Warner Books, 1987).

5. See C. K. Waddington, *OR in World War II: Operational Research Against the U-Boat* (London: Elek Science, 1973). There were, of course, many earlier anticipations of game theory: books and papers analyzing specific games like checkers and poker, monographs introducing concepts that were later incorporated into game theory, and books on real-world problems that reveal a style of thinking similar to that of game theory.

6. See Lester Thurow, *Zero-Sum Society* (New York: Basic Books, 1980).

7. The government does impose antitrust laws and other regulations, but

these are only a small portion of the rules by which business is played, and even these can be changed.

2. Co-opetition

1. Henry Ford, for his part, set up his own production studio to film 'Good Miles' movie clips, which were shown in the cinemas and whetted people's appetite for roads. For this history see Drake Hokanson, *The Lincoln Highway: Main Street Across America* (Iowa City: University of Iowa Press, 1989).

2. Other complements they give away. *La Centrale* offers free legal services to readers who discover that an advertiser has misrepresented a car. The result: honest ads and a better magazine.

3. Quoted in *Fortune*, July 10, 1995, cover. *Only the Paranoid Survive* is also the title of Grove's new book (HarperCollinsPublishers 1997).

4. Ibid., pp. 90–91.

5. The IBM OS/2, introduced in 1987, was the first PC-based 32-bit operating system, but it never took off. Apple Macintosh's System 7 was a second 32-bit operating system, but since it didn't use Intel chips, that didn't help Intel either.

6. *Wall Street Journal*, June 12, 1995, p. B3.

7. Because ISDN lines were so slow in being adopted, some took 'ISDN' to stand for 'It Still Does Nothing.'

8. *Wall Street Journal*, October 26, 1995, p. B8. Phone companies also benefit from a move to ISDN, because it helps fend off the challenge from wireless communication. With digital compression techniques and additional spectrum allocated by the FCC, there will soon be enough wireless voice capacity to send prices downward, threatening the market for the line carriers. But if phone companies can get people to pay a little more to make video calls using ProShare, then wireless is no longer a threat. The amount of capacity needed is so large that carrying video signals by wireless is currently impractical.

9. From now on, we will use 'product' to cover both products and services.

10. 'Revisiting Rationalization of America's Defense Industrial Base,' presentation to Aerospace Industries Association Human Resources Council, October 27, 1992.

11. The exercise of drawing one for the university, our own business, brought home to us how little of modern management thinking has made its way into universities. Students as customers? Working together with donors as partners? It sounds provocative, even heretical. Yet many universities need to start thinking this way. State schools

without large endowments seem to be ahead of private schools in being responsive to students, parents, and legislatures. But, overall, there is a growing public resistance to ever-rising college tuition and a growing perception that universities are not well managed. The state of higher education could well be the next big public debate in the United States, after health-care reform. We believe it would further that debate to think about universities as businesses – with customers, suppliers, competitors, and complementors, like any other business.

12. Although they can be argued to be complementors, too. Giving an exec ed program in a company helps drum up interest in having the company send its executives to a school program. It also makes the company more likely to hire the school's students.

13. Cynics might say that's why information technology is taking so long to come to the college campus.

14. American is also a supplier of MIS to other airlines. It sells yield management and loyalty management skills to other airlines and to other businesses, such as hotels and car rental agencies.

15. Gary Hamel and C. K. Prahalad, *Competing for the Future* (Boston: Harvard Business School Publishing, 1994).

16. The sales data come from Harold Vogel, *Entertainment Industry Economics* (Cambridge: Cambridge University Press, 1995). They are wholesale numbers for domestic and foreign sales.

17. Mary Westheimer, *Publishers Weekly*, August 28, 1995, p. 35.

18. Here, Mary Westheimer was referring back to the expansive view of UCLA marketing professor Ed Gottlieb: 'First you have to create an appetite for books; then people will buy them.'

19. Mary Westheimer, *Publishers Weekly*, August 28, 1995, p. 35.

20. *Wall Street Journal*, October 31, 1995, p. A1.

21. Harvard Business School professor Michael Porter discusses clustering effects in *The Competitive Advantage of Nations* (New York: Free Press, 1990).

22. Stephen Covey makes this point. See his book *The Seven Habits of Highly Effective People* (New York: Simon & Schuster, 1990), pp. 209–10.

3. Game Theory

1. There are many books on game theory and its applications. Perhaps we're a little biased, but we think a good starting point is Avinash Dixit and Barry Nalebuff, *Thinking Strategically: The Competitive Edge in Business, Politics, and Everyday Life* (New York: W. W. Norton, 1992). A classic that applies game theory to competitive situations is Thomas Schelling, *The Strategy of Conflict* (Cambridge: Harvard University Press, 1960). Of course, game theory also illuminates cooperative behavior; see Robert

Axelrod, *The Evolution of Cooperation* (New York: Basic Books, 1984). The following books apply game theory to evolutionary biology, law, politics, even the bible: Richard Dawkins, *The Selfish Gene* (New York: Oxford University Press, 1976); Douglas Baird, Robert Gertner, and Randal Picker, *Game Theory and the Law* (Cambridge: Harvard University Press, 1994); William Riker, *The Art of Political Manipulation* (New Haven: Yale University Press, 1986); Steven Brams, *Biblical Games: A Strategic Analysis of Stories in the Old Testament* (Cambridge: MIT Press, 1980). For anyone interested in delving further into the mathematical theory, we recommend Martin Osborne and Ariel Rubinstein, *A Course in Game Theory* (Cambridge: MIT Press, 1994) and Roger Myerson, *Game Theory* (Cambridge: Harvard University Press, 1991).

2. See analysis in *BusinessWeek*, November 20, 1995, p. 54, and *New York Times*, November 7, 1995, p. B9.

3. See Adam Brandenburger and Harborne Stuart, 'Value-Based Business Strategy,' *Journal of Economics & Management Strategy*, Spring 1996.

4. This was predictable. Why didn't the studio write a long-term contract, locking in Macaulay for a multimovie deal? In fact, it did, and Macaulay renegotiated the deal. Historically, movie studios had locked in actors through long-term – even lifetime – contracts. But the contracts were often unenforceable. Marilyn Monroe, for example, was signed to 20th Century-Fox indefinitely for $50,000 a picture. In 1954 she went on strike. Fox suspended Monroe for contract violation, but she held out, the public sided with her, and Fox came up with a new, more generous arrangement. See Donald Spoto, *Marilyn Monroe: The Biography* (New York: HarperCollins, 1993).

5. 'The Domino Effect: Of Foxes, Printers, and Prices,' *Channelmarker Letter*, vol. 2, no. 6 (December 1990), pp. 1–7.

6. James Charlton, ed., *The Executive's Quotation Book* (New York: St. Martin's Press, 1983).

7. This is not a new word. According to *Webster's Third New International Dictionary*, 'allocentric' means 'having one's interest and attention centered on other persons.'

8. Roger Fisher and William Ury, *Getting to Yes* (New York: Penguin, 1981), p. 23.

4. Players

1. Some information in the following story is drawn from 'Bitter Competition: The Holland Sweetener Company versus NutraSweet,' Harvard Business School Publishing, 9-794-079 to 9-794-085, 1993.

2. *Wall Street Journal*, May 7, 1987, p. 36.

3. *Food & Beverage Marketing*, March 1992, p. 36.
4. Pepsi was first to use 100 percent aspartame in its diet beverages and took advantage of its head start over Coke to promote Diet Pepsi against Diet Coke.
5. This comes from Monsanto's annual report. Even competitors acknowledged NutraSweet's overwhelming brand recognition. 'Alberto-Culver [maker of Sugar Twin Plus] research showed about 95% of sweetener users knew what NutraSweet was, while only 10% were familiar with the generic term aspartame, according to [their product manager] Jeff Clark' (*Advertising Age*, September 20, 1993).
6. For more details on this event, see Harvard University John F. Kennedy School of Government case 'Gainesville Regional Utilities,' written by Jose A. Gomez-Ibanez.
7. Confirmed in telephone conversation.
8. *Richmond Times-Dispatch*, March 7, 1993, p. E-1.
9. Ibid.
10. 'For Craig McCaw, It's Do-or-Die Time,' *BusinessWeek*, December 4, 1989.
11. At the same time, McCaw approached another cellular company, Metromedia, and offered a $1.9-billion bid for its half of the New York cellular license. LIN owned the other half and had a right of first refusal on any sale by Metromedia. Now LIN would either have to pay out $1.9 billion or share the property with McCaw. Either way, BellSouth would find LIN a less attractive prize. McCaw also lobbied Congress to pass legislation limiting cellular purchases by Baby Bells.
12. McCaw paid $26.5 million to LARCC (Los Angeles Radio Common Carrier). This was a joint venture between McCaw and BellSouth which was 85 percent owned by BellSouth. Since McCaw did not get any additional equity for his investment, it was in essence a $22.5-million payment to BellSouth.
13. Instead of ratcheting the fees down to a reasonable level, the courts disallow them entirely.
14. A personal conversation in which the executive asked to remain anonymous.
15. Courts reasoned that securities laws and antitrust laws deal with different issues, have different remedies, and that collusive bidding in corporate takeovers fits more properly under securities law. Courts have felt that if securities law did not preempt antitrust law, antitrust law would de facto preempt securities law. Antitrust laws have more generous remedies (treble damages) and more generous provisions regarding statutes of limitations and attorneys' fees. If not preempted, they

would be used almost exclusively, thereby nullifying the congressional intent in passing special securities laws. For more on this point, see R. Preston McAfee et al., 'Collusive Bidding in Hostile Takeovers,' *Journal of Economics and Management Studies*, Winter 1993, pp. 466–74.

16. Remember Groucho Marx: he didn't want to join any club that would have him as a member.

17. You can, in principle, think about bringing in an extended family of players – customers to your customers, suppliers to your suppliers, and so on. All the ideas of this chapter apply. However, we don't think this is the first order of business. If it's hard bringing in customers, it's that much harder bringing in customers to your customers. Moreover, the benefits to you are that much more diluted.

18. Some of the information in the following story is drawn from 'Harnischfeger Industries: Portal Cranes,' Harvard Business School Publishing, 9-391-130, 1991.

19. *Washington Post*, May 23, 1995, p. Cl.

20. *Health Alliance Alert*, May 26, 1995.

21. *Business Insurance*, May 29, 1995.

22. Charles Blanksteen, managing director of William Mercer, the consulting firm advising the coalition, explained: 'We wanted to understand the system behind the decisions that were made' (*Business and Health*, July 25, 1995).

23. *Business and Health*, July 1995.

24. And that's a real problem for brand-name goods. Michael Treacy discussed this issue in a 1995 talk to the Minneapolis Masters Forum.

25. For an argument promoting favorable treatment toward buying coalitions, see Jonathan M. Jacobson and Gary J. Dorman, 'Joint Purchasing, Monopsony, and Antitrust,' *Antitrust Bulletin*, Spring 1991, pp. 1–79.

26. Some of the information in the following story is drawn from 'Power Play (C): 3DO in 32-Bit Video Games,' Harvard Business School Publishing, 9-795-104, 1995.

27. 'Ego Trip,' *Marketing Computers*, April 1994, p. 18.

28. Personal interview, August 23, 1994.

29. '3DO Faces Revolt by Game Developers over Fee to Cut Manufacturers' Losses,' *Wall Street Journal*, October 24, 1994, p. B3.

30. Matsushita had 15 percent ownership of 3DO. That helped, but it wasn't enough.

31. *Fortune*, July 10, 1995, p. 20.

32. This analogy comes from Harvard Business School professor Michael Porter.

33. Tom Peters, *The Tom Peters Seminar: Crazy Times Call for Crazy Organizations* (New York: Vintage, 1994, p. 52).
34. Stephen Goddard, *Getting There: The Epic Struggle between Road and Rail in the American Century* (New York: Basic Books, 1994), chapter 7.
35. The problem is that the company that exits won't be around to reap any of the benefits. That's why it makes sense for one company in the industry to buy another, essentially paying it to exit the game.

5. Added Values

1. 'Can Nintendo Keep Winning?' *Fortune*, November 5, 1990, p. 131. However, as of 1995, Sony had the highest market capitalization.
2. David Sheff, *Game Over: How Nintendo Conquered the World* (New York: Vintage Books, 1993), p. 71. We recommend this book highly; it helped us greatly in telling the Nintendo story.
3. Some of the information in the following story is drawn from 'Power Play (A): Nintendo in 8-Bit Video Games,' Harvard Business School Publishing, 9-795-102, 1995.
4. Ibid, p. 14.
5. In an interview, Trip Hawkins, the founder of Electronic Arts, talked about his delay in getting into video games:

   It was my biggest mistake since starting the company ... Everyone figured [Nintendo] would be a Cabbage Patch doll kind of thing – that it would hold up for another year and then go the way of Atari and Coleco and the other video game systems that had disappeared ... There was another factor that kept us on the sidelines ... If you made software for Nintendo, you were restricted from moving to other video game machines for two years ... We didn't want to put all our eggs in one basket. But after a while there were so many Nintendos out that it was a moot point. (*Upside*, August/September 1990, p. 48)

6. 'Nintendo Paces Videogames: Attention Turns to Adults and New Product Tie-Ins,' *Advertising Age*, January 30, 1989, p. 24; 'Marketer of the Year,' *Adweek*, November 27, 1989, p. 15. A temporary worldwide chip shortage had something to do with the cartridge shortage. But there was more to it than that.
7. See Gary Becker, 'A Note on Restaurant Pricing and Other Examples of Social Influences on Price,' *Journal of Political Economy*, 1991, pp. 1109–16.
8. According to a 1990 'Q' survey. See *USA Weekend*, July 22, 1990, p. 14.

9. See 'Will Justice Department Probe Nintendo?' HFD – The Weekly Home Furnishings Newspaper, vol. 63, no. 51, p. 103.

10. Barron's, December 23, 1991.

11. On a separate issue, Nintendo made a settlement with the Federal Trade Commission in which it agreed to stop requiring retailers to adhere to a minimum price for the game console. Furthermore, Nintendo would give previous buyers a $5-off coupon toward future purchases of Nintendo game cartridges.

12. For more on the story of DeBeers and diamonds, see Debra Spar, The Cooperative Edge: The Internal Politics of International Cartels (Ithaca, NY: Cornell University Press, 1994), pp. 39–97.

13. See 'DeBeers Consolidated Mines Ltd. (A),' Harvard Business School Publishing, 9-391-076, 1990.

14. Wall Street Journal, October 31, 1994.

15. Because of capital adequacy requirements, declining profits actually lead to less capacity in the insurance business.

16. Personal conversation with Bob Cozzi.

17. New York Times, May 16, 1993.

18. Personal conversation with Bob Cozzi.

19. Michael Porter and Claas van der Linde, 'Green and Competitive: Ending the Stalemate,' Harvard Business Review, September/October 1995, pp. 120–34.

20. Truth be told, while we were busy writing this book, our friend Gus Stuart went to Club Med and vacationed for us. In addition to his reports, our data come from student papers and 'Club Med (A) & (B),' Harvard Business School Publishing, 9-687-046 and 9-687-047, 1986.

21. Many of the facts in this story are taken from the New York Times, August 19, 1995, p. 7.

22. Ibid.

23. Ibid.

24. Ibid.

25. Some of the information in the following story is drawn from 'The Free-Rider Problem: Airline Frequent-Flyer Programs,' Harvard Business School Publishing, 9-794-106, 1994.

26. Los Angeles Times, June 8, 1986, section 4. (PS – Who buys a sports car for fuel efficiency?)

27. Initially, American's offer was only to qualified frequent flyers. Delta then matched for people who bought their ticket with an American Express card. Then Eastern, United, and everyone matched, and all the restrictions disappeared.

28. Here's the calculation: A total of 1.2 trillion miles, with 25,000

miles required for a free flight, translates to 48 million free tickets. At 500 people on a 747, that's nearly 100,000 roundtrips of a full 747.

29. National Car Rental has savings bonds as one of its reward options.

30. Indeed, you don't want to give away your product to people who value it below what it costs you to provide it. In such cases, it's more cost-effective to give people cash.

31. Why don't the long-distance phone companies give out free voice mail, call waiting, and three-way calling as a reward to loyal customers? The problem is that these services are currently provided by the local, not the long-distance, carrier. And the local carrier gets paid an access charge by the long-distance carrier whether or not a call is completed. Even so, the long-distance carriers might want to buy these services in bulk from the local carriers and give them to their customers.

32. Sharon Oster, *Modern Competitive Analysis*, 2nd ed. (New York: Oxford University Press, 1994), p. 12.

33. Bruce Henderson, 'The Origin of Strategy,' in C. Montgomery and M. Porter, eds., *Strategy: Seeking and Securing Competitive Advantage* (Boston: Harvard Business School Press, 1991), pp. 3–4. The biologist's description of Gause's principle is: 'Two species with identical ecologies can't live together in the same place at the same time' (E. R. Pianka, *Evolutionary Ecology*, 4th ed. [New York: Harper & Row, 1988], p. 221).

34. See John Kay, *The Foundations of Corporate Success: How Business Strategies Add Value* (London: Oxford University Press, 1993), reviewed in *The Economist*, April 17, 1993, p. 65.

35. See Richard D'Aveni, *Hypercompetition* (New York: Free Press, 1994).

36. See David Collis, 'Understanding Competitive Advantage: The Role of Positioning, Sustainability, and Capabilities,' Harvard Business School working paper, 1995.

37. Some of the information in the following story is drawn from 'Minnetonka Corporation: From Softsoap to Eternity,' Harvard Business School Publishing, 9-795-163, 1995.

38. 'Softsoaping P&G,' *Forbes*, February 18, 1990, p. 91.

39. See Hirotaka Takeuchi and Ikujiro Nonaka, 'The New New Product Development Game,' *Harvard Business Review*, January/February 1986, pp. 137–46.

40. Our analysis of IBM's mistakes in the PC business was inspired by the illuminating discussion in Neil B. Niman, 'Lesson 3: Defending Yourself When Creating a Standard,' in *Standards: Strategic Lessons from the Computer Industry* (manuscript, Whittemore School of Business, University of New Hampshire), especially pp. 2–5. Niman sums up the turn of

events as follows: '[IBM] nurtured cooperation, only to discover that having made it to the dance, their partner went off courting others, leaving them all alone.'

41. See Paul Carroll, *Big Blues: The Unmaking of IBM* (New York: Crown, 1993), pp. 119, 131. Bill Gates discussed his one-time willingness to sell 30 percent of Microsoft to IBM in an interview in *Computer World*, May 24, 1993, p. 123. Of course, if IBM had owned a large stake in Microsoft over the last decade, Microsoft might not have been worth nearly $60 billion by 1996.

6. Rules

1. Richard H. Rovere, *Senator Joe McCarthy* (New York: Harcourt, Brace & World, 1959), p. 65. Harold J. Laski (1893–1950) was a controversial English intellectual who greatly influenced the British socialist movement from World War I on.

2. The term 'most-favored-nation clause' reflects its use in international trade where all countries so designated are given the lowest tariffs of any trading partners.

3. If Adam would have had to make the same concession later on anyway in order to close a deal in one of the subsequent negotiations, then it doesn't cost two dollars. However, at the time of the first few negotiations, Adam doesn't know what he'll have to do later on. Since Adam can never go back, agreeing to a new low price has two costs for Adam. First, he has to give the same deal to Tarun. Second, the more generous Adam is in the earlier rounds, the harder it becomes to use Tarun as a foil in the later rounds.

4. Originally, Congress passed the Communications Act of 1934, which guaranteed members 'lowest unit cost' (LUC) advertising from broadcasters. The act was amended by the Federal Election Campaign Act of 1971, which modified some of the original provisions regarding the period that candidates can purchase LUC advertising and other items, such as preemption of ads.

5. *Fortune*, December 27, 1993, p. 120.

6. Fiona Scott Morton, 'The Strategic Response by Pharmaceutical Firms to the Medicaid Most-Favored-Customer Rules,' *RAND Journal of Economics*, 28 (2), 1997.

7. *Fortune*, December 27, 1993, p. 120.

8. The traditional story, which has Cortés burning his ships, is in W. H. Prescott, *The History of the Conquest of Mexico*, vol. 1 (London: Gibbings & Co. [1843], 1896), chapter 8. For an account based on more modern research, see Hugh Thomas, *Conquest: Montezuma, Cortes and the Fall of Old*

Mexico (New York: Simon & Schuster, 1993), pp. 222−24. Thomas says that the legend of burning the ships arises from an error, where a reference to boats breaking (Spanish *quebrando*) in a contemporary document was misread as burning (*quemando*). For more on the strategic analysis of this story, see Avinash Dixit and Barry Nalebuff, *Thinking Strategically: The Competitive Edge in Business, Politics, and Everyday Life* (New York: W. W. Norton, 1992), and Richard Luecke, *Scuttle Your Ships Before Advancing* (New York: Oxford University Press, 1994).

9. *Wall Street Journal*, January 28, 1994.

10. The problem is that these contracts are hard to enforce when the information isn't public. For example, Gordie Howe, one of the great hockey players of all time, was reportedly told that he had the highest salary in hockey but that he shouldn't talk about it for fear of making other players jealous. Only later did he discover the real reason that management didn't want him talking about his salary − it wasn't the highest.

11. Salary caps in some of these sports create an extra layer of complexity which further changes the incentive to start bidding wars.

12. We are indebted to UCLA professor Sushil Bikhchandani for suggesting this application of the Card Game.

13. In the United States, many consumer electronics stores claim to have the same policy, but they don't always have the lowest prices. Sometimes these stores will use the proliferation of model numbers and unimportant differences as a reason not to match someone else's lower price. Other times the stores will exclude certain types of retailers (such as price clubs) from the universe of competitors they are willing to match.

14. In February 1993 GM launched the Gold card, raising the annual limit to $1,000 and the seven-year ceiling to $7,000.

15. Data from SMR Research Corp., Budd Lake, New Jersey. Also see 'The Strategic Alliance That Produced the GM Card,' *Direct Marketing Magazine*, September 1993, p. 64.

16. In technical terms, demand for cars becomes less 'elastic' − less sensitive to price.

17. This turn of phrase is due to our colleague, Dan Raff.

18. *BusinessWeek*, August 1, 1994, pp. 28−29.

19. *BusinessWeek*, February 13, 1995, p. 40.

20. The program saves marketing costs and lowers credit-card churn and its associated costs. It may also get people to buy cards sooner, which would raise GM's added value somewhat.

21. Telephone interview, December 16, 1994.

22. For an interesting discussion of the US antitrust stance toward so-called facilitating practices (in particular, some of the rules we've been looking at), see Michael Weiner, 'Distinguishing the Legitimate from the Unlawful,' *Antitrust*, Summer 1993, pp. 22–25. See also the pioneering analysis of Georgetown University professor Steven Salop, 'Practices That (Credibly) Facilitate Oligopoly Co-ordination,' in *New Developments in the Analysis of Market Structure*, J. Stiglitz and G. Mathewson, eds. (Cambridge: MIT Press, 1986).

23. Federal Trade Commission Decision, January 1, 1983, to June 30, 1983, vol. 101, pp. 657, 683.

24. In a larger sense, it was all a moot decision. The product in question, lead-based fuel additives, was headed for extinction with the end of leaded gas. By the end of 1985 Ethyl had exited the business, and Du Pont was down to one plant in New Jersey.

25. For example, companies can give volume discounts only to the extent that it is less expensive to serve large customers. The law itself is anticonsumer. It was designed to protect small corner grocery stores from the large supermarkets.

26. Sec. 2(b) of the Clayton Act, 38 Stat. 730 (1914), as amended, 15 USCA secs. 12–27 (1977).

7. Tactics

1. *Fortune*, June 13, 1994, p. 84. Also see R. Garda and M. Marn, 'Price Wars,' *McKinsey Quarterly*, no. 3, 1993, pp. 87–100.

2. Actually, convincing the publisher that the book will come in on time may well be impossible. This is because authors almost always underestimate how long it will take to write a book: if they were more realistic about the time needed, they would never agree to do it. (As we write this, our book is one month late. Well, really two months late, but who's counting?)

3. Richard Dawkins, *The Selfish Gene* (New York: Oxford University Press, 1976), p. 171.

4. Ibid., p. 172.

5. *New York Times*, July 4, 1994, p. 39.

6. *Financial Times*, August 23, 1995.

7. Peter Robinson, *Snapshots from Hell: The Making of an MBA* (New York: Warner Books, 1994).

8. The standard royalty rate is 10 percent on the first five thousand copies, 12.5 percent on the next five thousand copies, and 15 percent thereafter.

9. This is why you'd rather not be hired by a committee. If no one person

is on the line for how you perform, it's harder to find a guardian angel.

10. An interesting question is to whom should the guarantee be paid, the sender or the receiver? Sometimes it's the sender who suffers a large cost if the package doesn't make it, while other times it's the receiver. While the two parties could in principle figure out how to split any award that is made, that might just add to the tension at the time of failure. To keep things simple, why not just pay each of the two parties $100 and leave it at that?

11. See, for example, Christopher Hart, 'The Power of Unconditional Service Guarantees,' *Harvard Business Review*, July/August 1988, pp. 54–62.

12. For details, see 'Gillette's Launch of the Sensor,' Harvard Business School Publishing, 9-792-028, 1991.

13. Arthur Conan Doyle, *The Complete Sherlock Holmes Short Stories* (London: John Murray, 1928), pp. 326–27.

14. There was a small silver lining. When a studio sells a movie in 'turnaround,' it typically sells the script for its cost (plus interest) and retains the rights to five percent of the net profits. In the case of ET, five percent of the net was probably worth more than all the profits from *Starman*.

15. See David Scharfstein and Jeremy Stein, 'Herd Behavior and Investment,' *American Economic Review*, June 1990, pp. 465–79.

16. 'Cracks in the Crystal Ball,' *Financial Times*, September 30, 1995, p. 19.

17. Robert H. Gertner and Geoffrey P. Miller, 'Settlement Escrows,' *Journal of Legal Studies*, vol. 24 (January 1995), pp. 87–122.

18. The mediator could easily be replaced by a simple computer program.

19. With a 20 percent discount rate, a company with a 10 percent growth rate should have twice the P/E of a company with flat growth – hence the difference in price.

20. A problem with these agreements is that they force the seller to bear a large amount of risk or illiquidity as well as make the seller dependent on the new owner's management decisions. This limits the circumstances in which they can be used.

21. Or if the components are collectively overpriced, they can run this transaction in reverse.

22. *Wall Street Journal*, October 6, 1995, p. B9C.

23. JT (Japan Telecom) and TWJ (Teleway Japan) can also be hooked up to this service.

24. For more information on Value Pricing, see 'American Airlines' Value Pricing,' (A) and (B), Harvard Business School Publishing, 9-594-001 and 9-594-019, 1993. The (A) case is the source of the Crandall quote.

25. 'AMR's Bid for Simpler Fares Takes Off,' *Wall Street Journal*, April 9, 1992, p. B1.
26. *South Florida Business Journal*, May 7, 1993. It was widely reported that Tisch made this claim, although it's hard to find an original source for it.
27. The act prohibited the cable companies from raising their rates to pay for retransmission consent. However, that prohibition was only for one year.
28. *Broadcasting and Cable*, May 17, 1993.
29. ABC was then owned by Capital Cities, which also owned the cable channel ESPN. ABC was paid 12 cents per subscriber for creating an ESPN spin-off, ESPN2. NBC owned cable channel CNBC and used its capabilities to create America's Talking, a new all-talk cable channel. (*New York Times*, August 26, 1993, p. D18.)
30. *New York Times*, September 28, 1993, p. D1.
31. For a postmortem discussion of who was right and wrong, see 'Intel or IBM: Who Do You Trust?' *PC Magazine*, February 7, 1995, p. 29.

8. Scope

1. Even if two games don't share a common player, they can be linked by a chain. If games A and B have a common player, and so do games B and C, then games A and C are also linked.
2. Some of the information in the following story is drawn from 'Power Play (B): Sega in 16-Bit Video Games,' Harvard Business School Publishing, 9-795-103, 1995.
3. 'The Charge of the Hedgehog,' *Forbes*, September 2, 1991, p. 42.
4. According to a 1993 'Q' survey.
5. 'Games Companies Play,' *Forbes*, October 25, 1993, p. 68.
6. 'Nintendo's Show of Strength,' *Dealerscope Merchandising*, February 1991, p. 15.
7. President's Message, Nintendo Co., Ltd., 1991 Annual Report.
8. At the new reduced price level, the original 8-bit NES was renamed My First Nintendo and marketed as an entry-level system.
9. Dorothy Leonard-Barton, *Wellsprings of Knowledge: Building and Sustaining the Sources of Innovation* (Boston: Harvard Business School Press, 1995), ch. 2.
10. The decision not to create competition in the new-product market is the mirror image of Ford Motor Credit's strategy of creating competition in a complements market.
11. Some of the information in the following story is drawn from 'Minnetonka Corporation: From Softsoap to Eternity,' Harvard Business School Publishing, 9-795-163, 1995.

12. Perhaps some pumps could be placed in the kitchen or guest room, where previously there had been no bars of soap. But in the bathroom, liquid soap would replace bar soap. If liquid soap wasn't pure cannibalization of bar soap, it was perhaps '99 and 44/100' percent pure cannibalization.

13. Only one major brand of bar soap, Procter & Gamble's Zest, was detergent-based.

14. The electronic publishers may well be the authors themselves.

15. Sega offered a $40 adapter that made it possible to play its older 8-bit games on the Genesis. It was ironic that Sega was the one to engineer compatibility, since Sega's 8-bit system had failed to establish a significant installed base.

16. Gary Jacobson, analyst at Kidder, Peabody, quoted in 'Nintendo Cools Off,' *Advertising Age*, December 10, 1990, p. 20.

17. 20th Century-Fox offers the deal 'Buy *FernGully* and get $5 off *Star Wars*.' Warner Bros. offers the deal 'Buy *Batman Forever* and get $5 off Fuji film or $3 off any of the MGM/UA James Bond movies.'

18. *Corpus Christi Caller Times*, October 8, 1993.

19. *Communications Daily*, October 13, 1993.

20. Ibid.

21. *Corpus Christi Caller Times*, October 19, 1993.

22. Ibid., November 17, 1993.

23. *Television Digest*, November 22, 1993.

24. Some of this information is from 'Bitter Competition: The Holland Sweetener Company versus NutraSweet,' Harvard Business School Publishing, 9-794-079 to 9-794-085, 1993.

25. In late 1993, one year after the US market opened up, Holland Sweetener brought an additional 1500 tonnes of capacity on line.

Index

Index